D1600851

SET THEORY

The Structure of Arithmetic

SET THEORY

The Structure of Arithmetic

Norman T. Hamilton
Joseph Landin

Dover Publications, Inc.
Mineola, New York

Bibliographical Note

This Dover edition, first published in 2018, is an unabridged republication of *Set Theory and the Structure of Arithmetic,* originally published in 1961 by Allyn and Bacon, Inc., Boston.

Library of Congress Cataloging-in-Publication Data

Names: Hamilton, Norman T. (Norman Tyson), 1927- author. | Landin, Joseph, author.
Title: Set theory : the structure of arithmetic / Norman T. Hamilton, Joseph Landin.
Description: Dover edition. | Mineola, New York : Dover Publications, Inc., 2018. | Originally published: Boston : Allyn and Bacon, Inc., 1961. | Includes index.
Identifiers: LCCN 2017052336| ISBN 9780486824727 | ISBN 0486824721
Subjects: LCSH: Set theory. | Arithmetic—Foundations.
Classification: LCC QA248 .H28 2018 | DDC 511.3/22dc—23
LC record available at https://lccn.loc.gov/201705233

Manufactured in the United States by LSC Communications
82472101   2018
www.doverpublications.com

CONTENTS

PREFACE

1 THE ELEMENTS OF THE THEORY OF SETS

1.1	Introduction	1
1.2	The Concept of Set	3
1.3	Constants	4
1.4	Variables and Equality	7
1.5	Some Basic Notations and Definitions	10
1.6	Subsets; Equality of Sets; The Empty Set	12
1.7	The First Theorem	17
1.8	A (Very) Brief Section on Logic	18
1.9	The Algebra of Sets	25
1.10	Remarks on Notation and Other Matters	34
1.11	Some Special Sets	38
1.12	Ordered Pairs	43
1.13	Cartesian Products, Relations	46

1.14 Functions (or Mappings)   49
1.15 Equivalence Relations and Partitions   63
1.16 Mathematical Systems   72

2 THE NATURAL NUMBERS

2.1 The Definition of the Natural Numbers   74
2.2 The Ordering of the Natural Numbers   89
2.3 Counting   98
2.4 Finite Sets   101
2.5 Addition and Multiplication   106
2.6 The Relations between Order, Addition and Multiplication   112
2.7 The Principle of Finite Induction, Again   115
2.8 Sequences   117
2.9 Recursive Definitions   120

3 THE INTEGERS AND THE RATIONAL NUMBERS

3.1 Introduction   132
3.2 Definition and Properties of the Integers   133
3.3 Number-Theoretic Properties of the Integers: Generalized Operations   147
3.4 The Rational Numbers   157
3.5 The Arithmetic of the Rational Numbers   161
3.6 Conclusion: Integral Domains and Quotient Fields   173

4 THE REAL NUMBERS

4.1 The Mysterious $\sqrt{2}$ 177
4.2 The Arithmetic of Sequences   180
4.3 Cantor Sequences   187
4.4 Null Sequences   194
4.5 The Real Numbers   199

5 THE DEEPER STUDY OF THE REAL NUMBERS

5.1   Ordered Fields   214
5.2   Relations between Ordered Fields and R, the Field of
       Rational Numbers   221
5.3   The Completeness of the Real Numbers   227
5.4   Roots of Real Numbers   239
5.5   More Theorems on Ordered and Complete, Ordered Fields   244
5.6   The Isomorphism of Complete, Ordered Fields   249
5.7   The Complex Numbers   253

PREFACE

This book—the first in a series of three volumes—evolved from lecture notes for a course intended primarily for high school mathematics teachers. The purposes of the course were, first, to answer the question "What is a number?" and, of greater importance, to provide a foundation for the study of abstract algebra, elementary Euclidean geometry and analysis. The second and third volumes in this series will deal respectively with some of the elements of abstract algebra and the study of elementary geometry.

The question "What is a number?" is usually ignored in the elementary school curriculum, and perhaps rightly. However, regardless of whether this question is best avoided, the feeling is becoming widespread that secondary, or even primary, school students should be taught early to recognize that numbers are abstract entities as distinguished from the concrete entities—marks on paper—which are used to denote them. Thus, "1," "$\frac{1}{3} + \frac{2}{3}$," "$\pi/\pi$," and "$2 \int_0^1 \times dx$" all denote the number one. If children are taught this concept, the teacher will then want to know whether these marks can be assigned denotations and, if so, what the denotations may be.

Many working mathematicians have come to hold that much of mathematics, including the classical number systems, can be best based on set theory. Certainly, the language and concepts of set theory have become indispensable to mathematicians as a vehicle for the communication of his ideas. Thus, it is natural to base everything upon set theory. Taking this point of view, one must start (Chapter 1) with enough of the rudiments of set theory upon which to build. The rest of the text leads the reader along a path starting with a construction of the natural number system and ending with a construction of the real numbers. En route the basic properties of the several number systems are developed. On finishing the text the reader should be prepared for first courses in abstract algebra and in real variables.

We have paid little attention to the logical foundations of set theory. We operate with naive, intuitive set theory, being careful to insure that all proofs are easily carried out within the framework of an adequately axiomatized set theory. The one exception that the expert will note is in Chapter 2 where it is casually asserted that N is a set. In axiomatic set theory the proof requires an axiom of infinity (e.g., the statement itself). Although the question of antinomies in set theory is not treated within this book, it has usually been raised at some point in the course and the students seem to enjoy a bit of discussion of the topic.

There are two decisions that anyone writing a text on this subject matter must make. For the natural numbers, he must choose between Peano's postulates and von Neumann's construction. For the real numbers, the question involves Dedekind cuts versus Cantor sequences. In each case, we have taken the second alternative. In the first case, we feel that the difficulties the student faces are about the same either way, provided there is no cheating with, say, recursive definition. Also, having adopted the von Neumann alternative, the instructor can, if time permits, mention the Peano postulates and point out that the class has, in effect, been given an existence proof for them. By this time, most students seem to appreciate the point. As for the question of Dedekind versus Cantor, we have perhaps adopted the more difficult alternative. However, many students do continue with a study of real variables and for this experience the Cantor sequences provide a better preparation.

We are greatly indebted to Professors Robert G. Bartle, Pierce W. Ketchum, Echo D. Pepper and Wilson M. Zaring of the Mathematics Department, University of Illinois, who have taught from various earlier drafts of this book and who have given us both useful criticisms of the text and the benefits of their classroom experiences. We also wish to thank Professors William W. Boone and Herbert E. Vaughan, of the same department, who gave valuable suggestions for Chapters 1 and 2. We owe a particular debt of gratitude to Professor Zaring for his detailed and careful comments on every aspect of the next-to-last draft.

Finally, we are grateful to the members of the Academic Year Institutes at the University of Illinois from the years 1957 to date who attended the courses in which earlier versions of this book were taught. Whatever pedagogical merits the book may possess are due to our attempts to meet the high standards of our colleagues who are dedicated teachers of mathematics in secondary schools and colleges.

Norman T. Hamilton
Joseph Landin

Urbana, Illinois

SET THEORY

The Structure of Arithmetic

1

THE ELEMENTS OF THE
THEORY OF SETS

1.1. INTRODUCTION

At a first and casual thought the word "set" fails to conjure up any familiar mental associations in the mind of a novice at the Theory of Sets. Yet, the set concept is so much a part of our culture and our daily lives that the language we speak contains many special words to denote particular kinds of sets. For instance:

> 1. A *herd* is a collection or *set* of cattle.
> 2. A *flock* of sheep is a *set* of sheep.
> 3. A *bevy* is a *set* of quail.
> 4. A *clutch* is a *set* of eggs in a nest.
> 5. A *legal code* is a *set* of laws.

Similarly, there is a *school* of fish, a *pride* of lions, a *brace* of ducks, a *moral code*, and so on.

In elementary mathematics the use of set-theoretic concepts occurs with great frequency, albeit in a hidden way. Consider a few examples from elementary algebra and geometry.

6. The solutions, 1 and 2, of the quadratic equation $x^2 - 3x + 2 = 0$ comprise the *set* of solutions of the given quadratic equation.

7. The locus of the equation $x^2 + y^2 = 1$, a circle, is the *set* of all points whose coordinates satisfy this equation.

8. In algebra school books we find statements such as:

$$\text{In general, } a(b + c) = ab + ac.$$

The meaning of this statement is that for every replacement of a, b, c by names of real (and, also, of complex) numbers the statement resulting from $a(b + c) = ab + ac$ is true. Thus, the "general" statement is a statement concerning the members of the *set* of all real (or complex) numbers.

The list of examples of the concealed use of the set concept in the statements of elementary mathematics can be extended indefinitely since *all of them are really statements concerning sets or about the totality of members of certain sets.* The use of set-theoretic language in mathematics has the advantages of clarity and precision in the communication of mathematical ideas. But if these were the only advantages, one might argue: "Clarity and precision can be obtained by care in speaking and writing ordinary English (or whatever language is used in the school) without bothering to develop a special language for this purpose." Although this thesis is debatable, we do not join the debate at this point. Our reason is that the use of set-theoretic concepts goes deeper than the introduction of clarifying terminology. Most mathematical disciplines can be regarded as branches of set theory. Thus the theory of sets provides a mechanism for unifying and simplifying substantial parts of mathematics. In the course of the present book, it will be seen that the few simple set-theoretic ideas presented in this chapter are adequate for the development of much of elementary arithmetic (Volume I), algebra (Volume II), and elementary (Euclidean) plane geometry (Volume III). The same few basic ideas of set theory will be used time and again in each of these disciplines. And every concept in each of the above-named disciplines will be expressed exclusively in terms of the concepts studied in this chapter.

Although the ideas presented in Chapter 1 are truly simple, they may appear strange to the uninitiated reader. He may find himself asking, "What does this have to do with the mathematics with which I am familiar?" The strangeness will disappear as he progresses further into the text. Its vanishing can be accelerated by constructing numerous examples of the concepts introduced. The connection between this chapter and the more familiar aspects of elementary mathematics will require time to expound.   Indeed, this is the subject matter of our book. We urge the reader to have a little patience and read on.

What prior knowledge is required to read this book?   In the strictest sense one need only know how to read carefully and to write; little previous mathematical experience is needed.   However, we shall, on occasion, rely upon the reader's acquaintance with some of the simplest facts of elementary arithmetic, algebra and geometry.   These facts will not be used directly in the development of the subjects under consideration.   Their sole uses will be to illustrate certain concepts, to motivate others and, in general, to act as a source of inspiration for what we do here.

This book should not be read as a novel or a newspaper; a sharp pencil and a pad of paper are essential tools for a comprehension of what follows.   Careful attention to details will be rewarded.

1.2.   THE CONCEPT OF SET

It is beyond the scope of this book to attempt a formal (axiomatic) development of set theory, and therefore we begin by describing the concept of set in a heuristic way.

By a *set* we mean any collection of objects; the nature of the objects is immaterial.   The important characteristic of all sets is this: Given any set and any object, then exactly one of the two following statements is true:

(a) The given object is a member of the given set.

(b) The given object is not a member of the given set.

The above description of the concept of set is by no means the last word on the subject.   However, it will suffice for all the purposes of this book.   A deeper study of the basic ideas of set theory usually requires an introduction such as the present one.   Moreover, it would take us in a direction different from our proposed course—the study of elementary arithmetic, algebra and geometry.

EXAMPLES

1.   The set of all men named "Sigmund Smith" residing in the United States at 1:00 P.M., June 22, 1802.

2. The set of all unicorns that are now living or have ever lived in the Western Hemisphere.

3. The set of all points in the coordinate plane on the graph of $x^2 + y^2 = 1$.

4. The set of all points in the coordinate plane on the graph of $|x| \geq 1$.

5. The set of all points in the coordinate plane common to the graphs of $x^2 + y^2 < 1$ and $x > 1$.

6. The set Z of all integers.

7. The set E of all even integers.

8. The set of all tenor frogs now living in the Mississippi River.

9. The set of all tenor frogs now living in the Mississippi River and of all points in the coordinate plane on the graph of $x > 1$.

Before continuing with the technicalities of set theory, a few preliminary ideas are required. These will be discussed in Sections 1.3 and 1.4.

1.3. CONSTANTS

No doubt the reader is aware that the language in which this book is written—American English—possesses many ambiguities. Were it not so, the familiar and occasionally amusing linguistic trick known as the "pun" would be a rare phenomenon. Although there is no objection to being funny, any mathematical text should resist strenuously all tendencies to ambiguity and confusion. We shall try to minimize such tendencies by describing carefully the uses of several crucial terms and expressions. Foremost among such terms are the words "constant," "variable" and "equals." These terms are familiar to the reader from his earliest study of high-school algebra. But our uses of these words may differ from those he is accustomed to. Therefore it is suggested that he read this section as well as Section 1.4 with care.

Definition 1. A *constant* is a proper name. In other words, a constant is a name of a particular thing. We say that a constant *names* or *denotes* the thing of which it is a name.

EXAMPLES

1. "Calvin Coolidge" is a constant. It is a name of a president of the United States.

2. "2" is a constant. It is a name of a mathematical object—a number—which will be described in detail in Chapter 2.

Of course, a given object may have different names, and so distinct constants may denote the same thing.

3. During his political life, Calvin Coolidge earned the sobriquet "Silent Cal," because of his extraordinary brevity of speech. Thus "Silent Cal" is a constant and denotes Calvin Coolidge.

4. The expressions "$1 + 1$" and "$-2 + 5 - \frac{8}{2} + \frac{6}{3} + 1$" are constants and both denote the number two.

It may come as a surprise that some constants are built of parts which are themselves constants. Thus "$2 + 1$" is a constant built of "2" and "1", both of which are constants. In ordinary English, there are analogous situations. For instance, the name "Sam Jones" is composed of the two names "Sam" and "Jones."

Constants which denote the same thing are *synonyms* of each other. "Calvin Coolidge" and "Silent Cal" are synonyms; similarly, "2" and "$1 + 1$" are synonyms. Observe that a sentence which is true remains true if it is altered by replacing a name by a synonym. Similarly, if the original sentence is false, then the sentence so altered is likewise false. For example, consider the paragraph

> Calvin Coolidge was the third president of the United States. Calvin Coolidge was also, at one time, a governor of the State of Massachusetts.

The first sentence is false and the second one is true. If "Calvin Coolidge" is replaced throughout by "Silent Cal," we obtain

> Silent Cal was the third president of the United States. Silent Cal was also, at one time, a governor of the State of Massachusetts.

Again, the first sentence is false, the second is true.

In ordinary, daily conversation it happens rarely, if at all, that a name of a thing, i.e., a constant, and the thing denoted are confused with each other. No one would mistake the *name* "Silent Cal" for the *person* who was the thirtieth president of the United States. In mathematical discourse, on the other hand, confusions between names and the things named do arise. It is not at all uncommon for the *constant* "2" to be regarded as the *number* two which it names. Let us make the convention that enclosing a name in quotation marks makes a name of the name so enclosed. To illustrate this convention, consider the expressions

$$\boxed{\text{Silent Cal}}$$

and

$$\boxed{\text{``Silent Cal''}}$$

written *inside* the two boxes. The expression inside the upper box is a name for the thirtieth president of the United States. The expression inside the lower box is a name for the expression inside the upper box. Similarly, the expression inside

$$\boxed{\text{`` ``Silent Cal'' ''}}$$

is a name for the expression inside the box printed five lines above. Now consider the sentence

Silent Cal was famous for his brevity of speech.

This sentence *mentions* (or, refers to) the thirtieth president of the United States but it *uses* the name "Silent Cal." The name "Silent Cal" occurs in the sentence, while the thirtieth president *in the flesh* is not sitting on the paper. The sentence

"Silent Cal" has nine letters

mentions a name, and it *uses* a name of the name mentioned, to wit " "Silent Cal"." In referring to, or mentioning, the name "Silent Cal," we no more put that name in the sentence than we put Calvin Coolidge himself into the sentence referring to the thirtieth president. Notice that the sentence

"Silent Cal" was famous for his brevity of speech

is not only false, but even downright silly. For it asserts that a *name* was famous for a property attributable only (as far as we know) to a person.

1.4.   VARIABLES AND EQUALITY

Variables occur in daily life as well as in mathematics. We may clarify their use by drawing upon experiences shared by many people, even non-mathematicians.

Official documents of one kind or another contain expressions such as

(1.1)   I, _____, do solemnly swear (or affirm) that . . .

What is the purpose of the "_____" in (1.1)? Obviously, it is intended to hold a place in which a name, i.e., a constant, may be inserted. The variable in mathematics plays exactly the same role as does the "_____" in (1.1); it holds a place in which constants may be inserted. However, devices such as a "_____" are clumsy for most mathematical purposes. Therefore, the mathematician uses an easily written symbol, such as a letter of some alphabet, as a place-holder for constants. The mathematician would write (1.1) as, say,

(1.2)   I, x, do solemnly swear (or affirm) that . . .

and the "x" is interpreted as holding a place in which a name may be inserted.

Definition 2. A *variable* is a symbol that holds a place for constants.

Suppose a variable occurs in a discussion. What are the constants that are permitted to replace it? Usually an agreement is made, in some manner, as to what constants are admissible as replacements for the variable. If an expression such as (1.1) (or (1.2)) occurs in an official document, the laws under which the document is prepared will specify the persons who may execute it. These, then, are the individuals who are entitled to replace the variable by their names. Thus, with this variable is associated a *set* of persons and the names of the persons in the set are the allowable replacements for the variable. In general:

> With each variable is associated a set; the names of the elements in the set are the permitted replacements for the given variable. The associated set is the *range* of the variable.

The range of a variable in a mathematical discussion is usually determined by the requirements of the problem under discussion.

Variables occur frequently together with certain expressions called *quantifiers*. As one might judge from the word itself, quantifiers deal with "how many." We use but two quantifiers and illustrate the first as follows:

Let x be a variable whose range is the set of all real numbers. Consider the sentence

(1.3) *For each* x, if x is not zero, then its square is positive.

The meaning of (1.3) is

> *For each* replacement of x by the name of a real number, if the number named is not zero, then its square is positive.

The quantifier used here is the expression "for each." Clearly, the intention is, when "for each" is used, to say something concerning each and every member of the range of the variable. For this reason, "for each" is called the *universal quantifier*. It is a common practice to use the expressions "for all" and "for every" as synonymous with "for each," and these three expressions will be used interchangeably in this text.

Observe that if in place of (1.3) we write

(1.4) For each y, if y is not zero, then its square is positive.

where the range of y is also the set of all real numbers, then the meanings of (1.3) and (1.4) are the same. Similarly, y can be replaced by z or some other suitably chosen symbol without any alteration of meaning. Such replacement allows us considerable freedom in the choice of symbols for variables.

The use of the second quantifier is illustrated by the sentence

(1.5) *There exists* an x such that x is greater than five and smaller than six

where the range of x is the set of all real numbers. The meaning of (1.5) is

> There is *at least one* replacement of x by the name of a real number such that the number named is greater than five and smaller than six.

The expression "there exists" is the *existential quantifier*. The expression "there is" is regarded as synonymous with "there exists." Again, the reader may observe that if the variable x is replaced throughout (1.5) by y or some other properly chosen symbol, the range being the same, then the meaning of the new sentence is the same as that of (1.5).

Definition 3. If an occurrence of a variable is accompanied by a quantifier that occurrence of the variable is *bound*; otherwise it is *free*.

In mathematical discourse, variables frequently occur as free variables. For instance, one finds discussions beginning with expressions such as

If x is a nonzero real number, then . . .

or, such as

Let x be a nonzero real number.   Then . . .

Many mathematicians regard such forms of expression as ones in which the entire discussion is understood to be preceded by a quantifier.   For example, in elementary algebra texts, one sees statements such as

Let x be a real number.   Then,

$$x + 2 = 2 + x.$$

This is to be interpreted as meaning:

For all real numbers, x, $x + 2 = 2 + x$.

The practice of beginning a discussion with "If x is . . ." or "Let x be . . .," i.e., the practice of using the variable as free, will be adopted in many places throughout this book.   Just which of the two quantifiers is intended to precede the discussion will always be clear from the context.   Therefore we shall not attempt to give any formal rules for supplying the missing quantifier.

We have said that letters are used as variables.   It will also happen that letters will occur as constants.   The contexts in which a letter occurs will make clear whether a constant or a variable is intended.

We conclude this section with a brief discussion of *equality*.   Suppose $x, y, z, \ldots$ are variables all having the same range.

Definition 4. The expression "$x = y$" means that x and y are the same object.   The symbol " $=$ " is called *equals*.   "$x \neq y$" means that x and y are not the same object.

For instance, "$2 + 2 = 4$" means that $2 + 2$ and 4 are the same number.   Similarly, "Euclid = Author of the 'Elements' " means that Euclid and Author of the 'Elements' are the same person.

Throughout, we assume the following:

I. For each x, $x = x$.   In words, equals is *reflexive*.
II. For each x and for each y, if $x = y$, then $y = x$.   (Equals is *symmetric*.)
III. For each x, for each y, and for each z, if $x = y$ and if $y = z$, then $x = z$.   (Equals is *transitive*.)

1.5.   SOME BASIC NOTATIONS AND DEFINITIONS

Definition 5. If an object x is a member of a set A, we say that x *is an element of A* and write

$$x \in A.$$

For instance, the integer 1 is an element of the set Z (Example 6, page 4); therefore we write $1 \in Z$.

If an object y is not an element of a set B, we write

$$y \notin B,$$

and say "y is not an element of B."   Thus $1 \notin E$, where E is the set of Example 7, page 4.

Now suppose that S is a set consisting only of the objects denoted by "a," "b," "c," "d."   We write

(1.6) $$S = \{a,b,c,d\};$$

thus, S and $\{a,b,c,d\}$ are the same set.   If we know the names of all the elements of a set, and if the objects in it are not too numerous, then (1.6) gives a convenient way of representing this set.

EXAMPLES

1. Suppose a geometry class consists of the students Dan Doe, Evelyn Earp, Jane Jones, Sam Small, Joe Zilch.   Then we write

    Geometry class = {Dan Doe, Evelyn Earp, Jane Jones, Sam Small, Joe Zilch}.

2. $\{0,1,2,3\}$ is the set consisting of the numbers 0, 1, 2 and 3.   In Chapter 2 this set will receive a simpler name.

The order in which the names of objects in a set are listed is immaterial.   Therefore we regard

{Sam Small, Jane Jones, Dan Doe, Joe Zilch, Evelyn Earp}

and

{Jane Jones, Joe Zilch, Evelyn Earp, Sam Small, Dan Doe}

etc., as being the same geometry class.   Similarly, the set of Example 2 above may also be denoted by " {0,1,2,3} ," " {0,3,1,2} ," etc.

On occasion one knows names for all the elements of a set, but the elements are too numerous for the names to be listed conveniently.   In such a case, one may use dots (. . .).   For instance, suppose the set T consists of all the integers beginning with 0 and ending with 4,257. Then one writes

$$T = \{0,1, \ldots , 4{,}257\}.$$

There will be another notation for sets, but it will, together with some questions on notation not yet raised, be deferred until Section 1.10.
Again we emphasize that the elements of a set may be of any nature whatsoever.   In particular, the elements of a set *may themselves be sets*.

EXAMPLES

1. Let F be the set of all families now residing in the town of Foosland.[1]   Thus the elements of F might be the Jones family, the Smith family, the Robertson family, etc., and we write

$F = \{$the Jones family, the Smith family, the Robertson family, . . .$\}$

the dots indicating the names of the families which could be secured from a town directory or by means of a house-to-house canvass. Each of the elements of F is a family, and each family is, in turn, a set of persons.   For instance, the Jones family might consist of the people Sam, Zelda, Joe; i.e., Jones = {Sam, Zelda, Joe}. But neither Sam Jones, nor Zelda Jones, nor Joe Jones is an element of F, since F is a set of families and none of these three persons is a family.

---

[1] A town in east-central Illinois.

2. The National League (denoted by "*N.L.*") can be defined as the set of teams consisting of the Giants (G), the Dodgers (D), etc. (Your local newspaper will supply the names of the remaining teams.) So

$$N.L. = \{G,D, \ldots\}.$$

In turn, each team is a set of players. If Zilch is a pitcher for the Dodgers, then Zilch $\in D$; but by definition of *N.L.*, Zilch $\notin N.L.$

3. If the National League were defined as consisting of all of its teams *and* all of its players, then

$$N.L. = \{G,D, \ldots, \text{Zilch, Brown}, \ldots\};$$

in this case we would have

$$\text{Zilch} \in D, D \in N.L., \text{ and also Zilch} \in N.L.$$

in contrast to Example **2.**

EXERCISES

1. Using Examples 1–9 (pages 3 and 4), name several sets whose elements are, in turn, sets.

2. Name a few sets whose elements are sets of sets.

1.6.   SUBSETS; EQUALITY OF SETS; THE EMPTY SET

A comparison of the sets Z and E (defined in Examples 6 and 7, respectively, page 4) yields the conclusion that E is a part of Z. How is this conclusion reached? We deduce it in the following way:

> Every element of E is an even integer (definition of E).
> Every even integer is certainly an integer.
> Hence every element of E is an integer.
> But Z is the set of all integers (definition of Z).
> Therefore every element of E is an element of Z.
> Thus, Z contains all the elements of E.

The relationship between E and Z illustrates the concept of *subset*.

Definition 6. Let A and B be sets. *A is a subset of B* means that every element of A is an element of B.   The symbol "$A \subset B$" is used to abbreviate the sentence, "A is a subset of B."   We also say "A is contained in B," "A is included in B."   The symbol "$B \supset A$" is defined as meaning the same as "$A \subset B$"; in words, "B contains A," "B includes A."

If we use the $\in$-notation, the definition of subset can be stated in the following brief and convenient way:

Definition 6'. *A is a subset of B* means: for all x, if $x \in A$, then $x \in B$.

EXERCISES

1. Among Examples 1–9 (pages 3 and 4), find those sets which are subsets of other sets in the list.

2. Name several examples of sets and subsets.

Under what condition can one say that a set A is not a subset of a set B?   Let us reason as follows:

(α) "A is a subset of B" means that "every element of A is an element of B."

(β) If A is not a subset of B, then the statement "every element of A is an element of B" must be false.   Hence the *negation* of "every element of A is an element of B" must be true.   Therefore our task is to determine what is the negation of "every element of A is an element of B."   An example may help.

Consider the statement, "Every Martian is a bug-eyed monster." This statement can be rephrased in terms of set theory in the following way: Let M be the set of all Martians and let BEM be the set of all bug-eyed monsters.   Then the assertion, "Every Martian is a bug-eyed monster," is expressed by

$$M \subset BEM.$$

Now, suppose it is not true that every Martian is a bug-eyed monster. In other words, suppose it is false that every Martian is a bug-eyed monster.   This means that at least one (and possibly more than one) Martian is not a bug-eyed monster.   That is to say, there is a (at least one) Martian who is not a bug-eyed monster.   Thus, the statement:

There is a $y \in M$ such that $y \notin BEM$

is the negation of

$$M \subset BEM.$$

Returning to the general situation, we see that if the negation of "A is a subset of B" is true, then

(γ) There is a z such that $z \in A$ and $z \notin B$.

On the other hand, if (γ) is true, then the definition of subset is violated and therefore A is not a subset of B. Consequently, (γ) is a characterization of "A is not a subset of B."

Definition 7. $A \not\subset B$ means A *is not a subset of B.* $A \not\subset B$ and $B \not\subset A$ have the same meaning.

EXERCISES

1. Prove that $A \subset A$ is true for every set A.

2. Prove: If $A = B$ then $A \subset B$ and $B \subset A$.

3. Prove: If $A \subset B$ and $B \subset C$ then $A \subset C$.

4. Prove: If $A \subset B$ and $A \not\subset C$ then $B \not\subset C$.

5. Is it true that if $A \subset B$ and $B \not\subset C$ then $A \not\subset C$? If this statement is false, give examples in which $A \subset B$ and $B \not\subset C$ are true but $A \not\subset C$ is false.

6. Give some examples which illustrate the differences among $\in$, $\subset$ and $=$.

At the beginning of Section 1.2, we said that our development of set theory would not be axiomatic. We are going to violate this promise by introducing, explicitly, just one axiom.

In Exercise 2, above, it was required to prove that if $A = B$ then $A \subset B$ and $B \subset A$. The axiom we require is the converse, namely:

The Axiom of Extensionality.   If A, B are sets and if $A \subset B$ and $B \subset A$ then $A = B$.

To illustrate this axiom, we consider several examples:

1. Let A be the set of all living United States citizens named "Samuel Snork" having at least one female ancester born outside the U. S. Let B be the set of all living U. S. citizens named "Samuel Snork" who are less than 2,500 years of age. It is evident that the sets A and B are not defined in the same way, yet there is no doubt that A and B are the same set, i.e., that $A = B$. What argument would convince us of the equality of A and B? The idea is to show that A and B are precisely the same set of elements. This can be done by showing that every element in A is also an element in B, and conversely, that every element in B is an element in A. Thus, let x be an element in A; then x is a living U. S. citizen named "Samuel Snork" who has at least one female ancestor born outside of the U. S. The mere fact that x is alive assures us that x is less than 2,500 years of age. Consequently, x *also satisfies all the criteria for membership in B.* Thus from

$$x \in A \text{ follows } x \in B$$

whence $A \subset B$. Conversely, from $x \in B$ follows $x \in A$, and therefore $B \subset A$. (Provide all the details of the argument.) From these considerations we may conclude, by the Axiom of Extensionality, that

$$A = B.$$

2. Let C be the set of all equilateral triangles and D the set of all equiangular triangles in the plane of elementary geometry (the Euclidean plane). The definitions of C and D are not identical, yet we are confident that $C = D$. Using Example 1 as a model, we may argue thus: Let $x \in C$. Then x is an equilateral triangle. By certain theorems of elementary geometry, we know that x is equiangular. Hence x satisfies the criteria for membership in D, and therefore from $x \in C$ follows $x \in D$, i.e.,

$$C \subset D.$$

Similarly (details?) $D \subset C$, and therefore, by the Axiom of Extensionality,

$$C = D.$$

EXERCISE

Give examples similar to 1 and 2.

A convenient device for picturing sets and relationships among them is provided by Venn diagrams. The idea is to represent sets by simple plane areas. Thus, if $A \subset B$, we can represent this situation diagrammatically in the following ways:

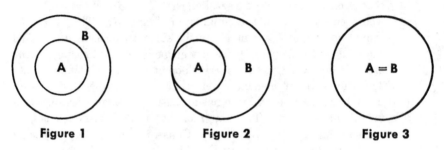

Figure 1   Figure 2   Figure 3

and so on. If $A \not\subset B$, then we have pictures such as

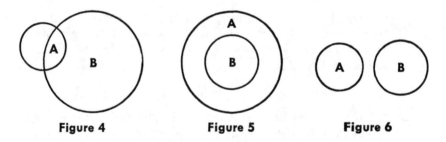

Figure 4   Figure 5   Figure 6

In each of the Figures 4–6, one sees that there is an element of A (i.e., a point) which is not an element (point) of B.

When we study the algebra of sets (Section 1.9) we shall find that these diagrams are helpful in verifying some relationships among sets, and in discovering others.

Among the Examples of Section 1.2, some, such as 2 and 5, may have raised honest doubts as to the seriousness of the writers of this text. Natural historians maintain that there is not now, nor has there ever been, a unicorn anywhere in the world. Therefore the set defined in 2 must contain no elements. The set defined in 5 also contains no elements.

These examples illustrate the concept of *empty (null, void)* set.

Definition 8. The *empty set*, denoted by "ϕ," is the set which contains no elements. If we use the $\in$-notation introduced earlier, we can say that ϕ is the set such that for all x, $x \notin \phi$.[2]

There is, on occasion, some confusion between the meanings of "empty set" and the word "nothing." Whatever meaning the reader may attribute to "nothing" the word is not one of the terms of our discipline; it will serve to avoid difficulty if we refrain from using it. As for "empty set," the only meaning allowed is the one given in Definition 8.

1.7.   THE FIRST THEOREM

We prove

Theorem 1. For each set A, $\phi \subset A$; in words, the empty set is a subset of every set.

We are going to prove the theorem by contradiction. (This is an *indirect proof*.) In outline, the procedure is:

1. Begin by assuming that the statement of the theorem is false.
2. If the given statement is false, then its negation must be true.
3. We prove that the negation leads to a contradiction of something known previously.
4. Consequently, the negation must be false.
5. Therefore the statement of the theorem is true.

Proof of Theorem 1

1. Assume that "For each set A, $\phi \subset A$," is false. This means that there is a (at least one) set, say B, for which the statement "$\phi \subset B$" is false. Thus, from the falsity of "For each set A, $\phi \subset A$" we infer that:
2. There is a set B such that $\phi \not\subset B$ (this is the negation of the theorem). Further, $\phi \not\subset B$ means (see (γ), page 14) that there is an element, x, such that $x \in \phi$ and $x \notin B$. Hence
3. $x \in \phi$. This contradicts the definition of ϕ.

---

[2] Although no range has been specified for the variable in Definition 8, the range of x is to be taken as the totality of all objects under consideration.

4. Consequently, statement 2 is false.
5. Therefore the theorem is true.

q.e.d.

We give a direct proof of Theorem 1 with the warning that it is a little harder to follow than the indirect proof.

Direct Proof of Theorem 1. We wish to show that each element $x \in \phi$ is also an element of A. But, since ϕ has no elements, it is certainly true that each element of ϕ is an element of A. Therefore $\phi \subset A$.

EXERCISES

1. Prove that every element of ϕ is a crocodile.

2. Prove that every element of ϕ is not a crocodile.

Definition 9. *A is a proper subset of B* means that $A \subset B$, and $A \neq B$. *A is an improper subset of B* means that $A = B$.

Before continuing with the development of set theory, we call the reader's attention to a few simple tools from logic which are used in proving theorems. Some of these tools have been used already in the foregoing.

1.8. A (VERY) BRIEF SECTION ON LOGIC

For the purposes of this book, a *sentence* is a declarative sentence which is either *true* or *false*, but not both. Given several sentences, there are, in logic and mathematics, standard ways of combining them to obtain new ones; the given sentences are called the *parts* of the new sentences. The purpose of this section is to provide and illustrate rules for determining the truth or falsehood of new sentences obtained in the standard ways when the truth or falsehood of the parts is known. To facilitate the discussion, let $p, q, r, \ldots$ be variables which are replaceable by sentences.

Definition of "or." A sentence

$$p \text{ or } q$$

is true if at least one of the two sentences (the parts) p, q is true; otherwise it is false. The sentence p *or* q is the *disjunction of p and q.*

EXAMPLES

1. The sentence, "17 is a number *or* three is a factor of 12," is true.

2. The sentence, "17 is a number *or* three is a purple cow," is true.

3. The sentence, "17 is not a number *or* three is a purple cow," is false.

A good "visual aid" to help one remember the definition of *or* is the so-called truth table for disjunction:

p	q	p or q
T	T	T
T	F	T
F	T	T
F	F	F

The table is interpreted thus:

The first line states that if the sentence p is true and if the sentence q is true, then p *or* q is true. The second line states that if the sentence p is true and if the sentence q is false, then p *or* q is true. And so on.

Rule I: The sentences

$$p \text{ or } q$$

and

$$q \text{ or } p$$

are both true or they are both false.

EXAMPLES

1. The sentences "17 is not a number *or* three is a factor of 12," and "Three is a factor of 12 *or* 17 is not a number," are both true.

2. The sentences "17 is not a number *or* three is a purple cow," and "Three is a purple cow *or* 17 is not a number," are both false.

Definition of "and." A sentence

$$p \text{ and } q$$

is true if the sentence p is true and the sentence q is true; otherwise it is false. The sentence, *p and q*, is the *conjunction of p and q*.

EXAMPLES

1. The sentence "17 is a number *and* three is a factor of 12," is true.

2. The sentence "17 is not a number *and* three is a factor of 12," is false.

3. The sentence "17 is a number *and* three is a purple cow," is false.

4. The sentence "17 is not a number *and* three is a purple cow," is false.

The truth table for conjunction is:

p	q	$p \text{ and } q$
T	T	T
T	F	F
F	T	F
F	F	F

The interpretation of this table is simple and is left to the reader.

Rule II: The sentences

$$p \text{ and } q,$$

and

$$q \text{ and } p,$$

are both true or they are both false.

Next, consider the sentence

(1.7) The moon is made of Liederkranz cheese.

Accepting the evidence of astronomy, this sentence is false.   If we prefix it with the words "It is false that" we obtain a new sentence

(1.8)   It is false that the moon is made of Liederkranz cheese.

This sentence is true.   It is also linguistically clumsy and so we convert it into the true sentence

(1.9)   The moon is not made of Liederkranz cheese.

The sentence (1.9) (or (1.8)) is the *negation* of (1.7).
     On the other hand, if we have a true sentence, say,

$$2 + 3 = 5,$$

then its negation,

$$\text{It is false that } 2 + 3 = 5,$$

or,

$$2 + 3 \neq 5$$

is false.

Definition of "not."   The negation of a false sentence is a true sentence; the negation of a true sentence is false.   The negation of a sentence p is written

$$not\text{-}p.$$

The truth table for negation is very simple:

p	$not\text{-}p$
T	F
F	T

     Two sentences built from the sentences p, q, r, . . . by means of *or*, *and*, and *not* and having the same truth values (i.e., both are true or both are false) regardless of the truth values of the parts p, q, r, . . . , are *equivalent*.   Thus, by Rule I, p *or* q and q *or* p are equivalent, and by Rule II, so are the sentences p *and* q and q *and* p.

Rule III: If a sentence is built from parts p, q, r, . . . by means of *or*, *and*, and *not*, and if any part is replaced by an equivalent sentence, then the result is a sentence equivalent with the original one.

EXAMPLES

1. The sentences (*p and q*) *or r* and (*q and p*) *or r* are equivalent, since the second is obtained from the first by replacing the part *p and q* by the equivalent *q and p*.

2. The sentences *p and q* and *p and not-(not-q)* are equivalent. This can be determined from the third and last columns of the truth table:

p	q	p and q	not-q	not-$(not$-$q)$	p and not-$(not$-$q)$
T	T	T	F	T	T
T	F	F	T	F	F
F	T	F	F	T	F
F	F	F	T	F	F

The second and fourth columns of the table also show that the sentences *q* and *not-(not-q)* are equivalent.

Two important rules (IV and V) deal with equivalents of negation.

Rule IV: The sentences *not-(p or q)* and *not-p and not-q* are equivalent.

A truth table demonstrates this equivalence at once:

p	q	p or q	not-$(p$ or $q)$	not-p	not-q	not-p and not-q
T	T	T	F	F	F	F
T	F	T	F	F	T	F
F	T	T	F	T	F	F
F	F	F	T	T	T	T

A comparison of the fourth and last columns shows that the sentences *not-(p or q)* and *not-p and not-q* are equivalent.

Rule V: The sentences *not-(p and q)* and *not-p or not-q* are equivalent.

EXERCISES

1. Construct truth tables exhibiting the equivalence of: (a) (*p and q*) *and r* and *p and* (*q and r*); (b) (*p or q*) *or r* and *p or* (*q or r*); (c) *not-(p and q)* and *not-p or not-q*; (d) *not-(not-p and not-q)* and *p or q*; (e) *p and* (*q or r*) and (*p and q*) *or* (*p and r*); (f) *p or* (*q and r*) and (*p or q*) *and* (*p or r*); (g) *p and* (*q and r*) and (*p and q*) *and* (*p and r*).

2. Are the sentences *not-(not-p or not-q)* and *p and q* equivalent? Are *not-(p and q)* and *not-p and not-q* equivalent? How about the sentences *not-(p or q)* and *not-p or not-q*?

Perhaps the only strange definition is the one dealing with *implication*.

Definition of "implies."   If the sentence p is true and the sentence q is false, then the sentence p *implies* q is false; otherwise it is true. p is the *antecedent* and q is the *consequent* in the sentence p *implies* q.

In place of the sentence p *implies* q one frequently writes:

If p then q.

It comes as no surprise that in a sentence p *or* q (or, p *and* q) the parts p and q are unrelated. But in ordinary usage, a statement to the effect that p *implies* q is usually taken to mean that p and q are related. A peculiar feature of our definition is that the antecedent and consequent may have no connection at all; yet this feature is vital to mathematical usage.

EXAMPLES

1. The sentence

If 17 is a number then three is a factor of 12,

is true because both the antecedent

17 is a number,

and the consequent

Three is a factor of 12,

are true.

2. "If 17 is a purple cow, then the reader of this sentence is a monkey's uncle," is true because both the antecedent and the consequent are false.

3. "If 17 is a purple cow, then three is a factor of 12," is true because the antecedent is false and the consequent is true.

4. "If 17 is a number, then the reader of this sentence is a monkey's uncle," is false because the antecedent is true and the consequent is false.

Note that in none of the examples 1–4 are the antecedent and consequent related.

EXERCISES

1. Make a truth table for the sentence p *implies* q.

2. Are the sentences p *implies* q and *not-q implies not-p* equivalent? Are the sentences p *implies* q and *not-p implies not-q* equivalent?

The final rule is:

Rule VI: (Modus Ponens) If the sentence p is true and if the sentence p *implies* q is true, then the sentence q is true.

EXAMPLES

1. Consider the sentences

T is an equilateral triangle.
If T is an equilateral triangle, then T is an equiangular triangle.

By the theorems of elementary geometry, we know that the second sentence is true. Hence, if T is indeed an equilateral triangle, it follows by Rule VI that T is equiangular.

2. Consider the sentences

Three is a purple cow.
If three is a purple cow, then 17 is a number.

The second sentence is true, the first is false. Hence we cannot deduce

17 is a number.

(Question: Does it follow that "17 is a number" is false?)

The above résumé of a small part of logic has been brief indeed, but it will suffice for many (but not all) of our purposes. These rules will be used as necessary, usually without referring to them. Wherever a new mode of reasoning is required, it will be explained at the point at which it is needed.

1.9.   THE ALGEBRA OF SETS

The term "algebra" usually reminds one of an enterprise involving numbers and operations, such as addition, multiplication, etc. The use of the term "algebra" in the present connection may seem far-fetched because, first of all, we are going to deal with sets and not with numbers. Furthermore, the operations we shall require are not the familiar ones of algebra. Nevertheless, the formalism to be developed here bears certain resemblances to the algebra we know.

Let A be the set of positive integers, B be the set of all integers less than eleven. What elements do the two sets have in common? Clearly these elements are the integers 1, 2, 3, 4, 5, 6, 7, 8, 9, 10. The set of integers common to both A and B is called the *intersection* of A and B.

Let C be the set of points in the coordinate plane on the graph of $x > -1$, D the set of points on the graph of $x < 1$. The points common to the two graphs are all the points and only those points (x,y) such that $-1 < x < 1$. Thus the *intersection* of C and D may be described as the graph of $|x| < 1$. (Draw a figure illustrating the intersection of C and D.)

Definition 10. The *intersection* of two sets A and B is the set of all elements common to both A and B. If we denote the intersection of A and B by "$A \cap B$" (read "A cap B") the definition can be stated more simply:

Definition 10′. The *intersection* of A and B is the set, $A \cap B$, of all elements x such that $x \in A$ and $x \in B$.

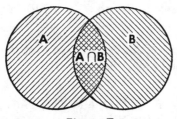

Figure 7

In Venn diagrams the intersection $A \cap B$ is pictured (Figure 7) as the cross-hatched area in which the two areas A, B overlap.

Theorem 2. $A \cap B = B \cap A$; in words, intersection is commutative.

proof: To use the Axiom of Extensionality, we must prove that

$$A \cap B \subset B \cap A \text{ and } B \cap A \subset A \cap B.$$

In order to prove that $A \cap B \subset B \cap A$ we show[3] that every element of $A \cap B$ is an element of $B \cap A$. Now, for each element $x \in A \cap B$ we know that $x \in A$ and $x \in B$ (Definition 10'). Hence, $x \in B$ and $x \in A$. Therefore (Definition 10') $x \in B \cap A$. In short, for all $x \in A \cap B$ we have proved $x \in B \cap A$. Consequently $A \cap B \subset B \cap A$ (Definition 6). In a similar fashion we can show that $B \cap A \subset A \cap B$. Therefore $A \cap B = B \cap A$.

EXERCISE

Supply the missing details in the above proof.

Now let A be the set of all positive integers, let B be the set of all integers less than eleven and let C be the set of even integers. Applying Definition 10' one sees that $A \cap (B \cap C) = \{2,4,6,8,10\}$ and also $(A \cap B) \cap C = \{2,4,6,8,10\}$. In a similar fashion, if the reader experiments with a variety of choices for sets A, B and C he will observe, in every case, it turns out that $(A \cap B) \cap C = A \cap (B \cap C)$. This leads us to conjecture

Theorem 3. For all sets A, B and C, $(A \cap B) \cap C = A \cap (B \cap C)$.

proof: The equality is established by means of the Axiom of Extensionality. Thus we shall prove that

$$(A \cap B) \cap C \subset A \cap (B \cap C) \text{ and } A \cap (B \cap C) \subset (A \cap B) \cap C.$$

For each $x \in (A \cap B) \cap C$ we have $x \in A \cap B$ and $x \in C$. But

[3] To avoid boring repetitions of the word "prove," the words "show," "verify," "establish" are used as synonyms for "prove."

if $x \in A \cap B$ then $x \in A$ and $x \in B$.   Thus from $x \in (A \cap B) \cap C$ we deduce $(x \in A$ and $x \in B)$ and $x \in C$.   This yields $x \in A$ and $(x \in B$ and $x \in C)$ whence $x \in A$ and $x \in B \cap C$.   Therefore $x \in A \cap (B \cap C)$ and so we have proved the first inclusion.

The strategy of proof of the second inclusion is the same as the foregoing and is left to the reader.

Definition 11. $A \cap B \cap C = (A \cap B) \cap C$.

By Theorem 3 we also have $A \cap B \cap C = A \cap (B \cap C)$.   In Figure 8, below, the shaded area illustrates $A \cap B \cap C$.

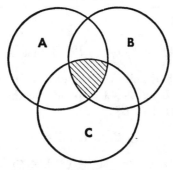

Figure 8

We consider another set-theoretic concept called the "union."   Let A be the set of all positive integers, B the set of all integers less than eleven.   The set consisting of all the elements of the given sets has as its elements the whole set of integers and this set is the *union of A and B*. Let C be the union of A and B.   We observe that C could be defined as the set of all integers x such that x is an element of at least one of the two sets A and B.   This definition of C is easily seen to agree with the one given before.

As a second example, let D be the set of all male animals in North and South America, E the set of all male animals in Europe and North America.   The union F of D and E consists of all male animals in Europe, North America and South America.   Note that F can also be defined as the set of all male animals x such that x is in at least one of the two sets D and E. This definition of the union of D and E yields the same set as before.

Finally, let G be the set of all tenor frogs in the Mississippi River, H the set of all coloratura toads in Lake Erie. In this case, even though the sets G and H have no elements in common, we may think of their union in the same way as in the preceding examples. The union K of G and H is the set of all animals x such that x is an element of at least one of the two sets G and H.

Definition 12. Let A, B be sets; the *union* of A and B, $A \cup B$ (read "A cup B"), is the set of all elements x such that $x \in A$ or $x \in B$.

EXERCISES

1. For each set A, what are $A \cup \phi$, $A \cap \phi$, $A \cup A$, $A \cap A$?

2. If A is the set of points in the coordinate plane on the graph of $x \geq -1$, and B is the set of points on the graph of $x \leq 1$, what are $A \cup B$ and $A \cap B$?

3. If $A \subset B$, prove $A \cup B = B$ and $A \cap B = A$.

4. For all A and B prove $A \subset A \cup B$, $A \cap B \subset A$. Prove that if $B \subset C$ then $A \cap B \subset A \cap C$.

5. Using Definition 11 as a model, define $A \cup B \cup C$. Prove $(A \cup B) \cup C = A \cup B \cup C = A \cup (B \cup C)$.

6. If $A \subset C$, $B \subset C$ then $A \cup B \subset C$; if $A \subset C$, $B \subset D$ then $A \cup B \subset C \cup D$.

In elementary algebra, we are familiar with the distributive law which states that for all real numbers, a, b, c, $a(b + c) = ab + ac$. In the algebra of sets, we have two distributive laws, namely

Theorem 4. For all sets A, B, C,

1. $A \cap (B \cup C) = (A \cap B) \cup (A \cap C)$;

2. $A \cup (B \cap C) = (A \cup B) \cap (A \cup C)$.

proof: To prove $A \cap (B \cup C) = (A \cap B) \cup (A \cap C)$ we show that $A \cap (B \cup C) \subset (A \cap B) \cup (A \cap C)$ and $(A \cap B) \cup (A \cap C) \subset A \cap (B \cup C)$.

If $x \in A \cap (B \cup C)$ then $x \in A$ and $x \in B \cup C$, so that $x \in B$ or $x \in C$. We make cases according as (a) $x \in B$, (b) $x \in C$.

case (a):   Here $x \in A$ and $x \in B$, hence $x \in A \cap B$. Since $(A \cap B) \subset (A \cap B) \cup (A \cap C)$ (Exercise 4, above), we deduce $x \in (A \cap B) \cup (A \cap C)$.

case (b):   Here $x \in A$ and $x \in C$. Exactly as above, $x \in A \cap C \subset (A \cap B) \cup (A \cap C)$.

In either case, if $x \in A \cap (B \cup C)$ we find $x \in (A \cap B) \cup (A \cap C)$; that is, $A \cap (B \cup C) \subset (A \cap B) \cup (A \cap C)$.

On the other hand, since $B \subset B \cup C$, we have $A \cap B \subset A \cap (B \cup C)$ (Exercise 4). Likewise, $C \subset B \cup C$; therefore $A \cap C \subset A \cap (B \cup C)$. Thus, both $A \cap B$ and $A \cap C$ are subsets of $A \cap (B \cup C)$. Using Exercise 6 (above), it follows that $(A \cap B) \cup (A \cap C) \subset A \cap (B \cap C)$.

We have proved the two inclusions needed to show that $A \cap (B \cup C) = (A \cap B) \cup (A \cap C)$.

We leave the proof of 2 as an exercise.

EXERCISE

Verify 1 and 2 for several triples of sets.

Occasionally it is convenient to regard $\cup$ and $\cap$ as analogues of $+$ and $\cdot$ respectively in algebra; indeed, because of the analogies, the formalism being developed here is called the "algebra of sets." Clearly the analogies are by no means exact, since there is no law of arithmetic corresponding to 2 of Theorem 4.[4]

A third set-theoretic concept is the "complement." Let A be the set of all fish in the Mississippi River, B the set of all catfish in the same river. Then $A - B$ is the set of all fish in the Mississippi which are not catfish. Thus $A - B$ is the set of all elements x such that $x \in A$

---

[4] The corresponding equation of algebra would be $a + (b \cdot c) = (a + b) \cdot (a + c)$ which is not valid for all real numbers a, b, c.

and $x \notin B$. In this example, we note that $B \subset A$. $A - B$ is the *complement* of B relative to A.

Let C be the set of all integers less than eleven, D the set of all positive integers. Then $C - D$ is the set of all integers ≤ 0. Again, $C - D$ can be obtained as the set of all elements x such that $x \in C$ and $x \notin D$ even though D is *not* a subset of C. Our definition is

Definition 13. For any sets A, B the *complement of B relative to A* is the set of all x such that $x \in A$ and $x \notin B$. It is denoted by "$A - B$."

The Venn diagrams below illustrate three different situations involving relative complements; the shaded areas represent $A - B$.

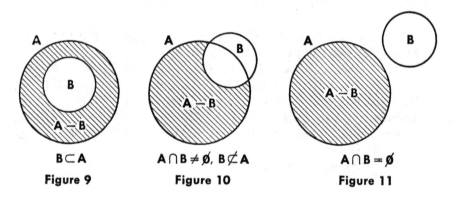

$B \subset A$	$A \cap B \neq \emptyset$, $B \not\subset A$	$A \cap B = \emptyset$
Figure 9	**Figure 10**	**Figure 11**

EXERCISE

Prove: For all sets A, B, $A - B \subset A$.

Theorem 5. For all sets E and A, $A \cap (E - A) = \phi$.

proof: We wish to show that $A \cap (E - A)$ contains no elements. Indeed, suppose $x \in A \cap (E - A)$; then $x \in A$ and $x \in E - A$ (Definition 10). But if $x \in E - A$ then $x \in E$ and $x \notin A$. Thus, if there were an element x in both A and $E - A$, this element would have the contradictory properties $x \in A$ and $x \notin A$. Therefore $A \cap (E - A) = \phi$.

q.e.d.

Theorem 6. For all sets E and A, $A \cup (E - A) = A \cup E$.

proof: We prove that $A \cup (E - A) \subset A \cup E$ and $A \cup E \subset A \cup (E - A)$.

The first inclusion is immediate. For, from $E - A \subset E$ and $A \subset A$, we deduce (Exercise 6, page 28) that $A \cup (E - A) \subset A \cup E$.

To prove the second inclusion, let x be an element in $A \cup E$; then $x \in A$ or $x \in E$. We distinguish two cases.

case 1: $x \in A$; then $x \in A \cup B$ for every set B (Definition 12); hence, in particular, $x \in A \cup (E - A)$.

case 2: $x \notin A$; in this case $x \in E$. We now have $x \in E$ and $x \notin A$ so that $x \in E - A$ (definition of $E - A$).

In short, for all x such that $x \in A \cup E$ we have $x \in A$ or $x \in E - A$, hence $x \in A \cup (E - A)$. Therefore $A \cup E \subset A \cup (E - A)$; with the inclusion obtained before, this yields the theorem.

q.e.d.

corollary: If $A \subset E$ then $A \cup (E - A) = E$.

proof: Exercise.

The next theorem is not at all surprising.

Theorem 7. For all sets E, $E - E = \phi$, $E - \phi = E$.

proof: Exercise.

Theorem 8. For all subsets A and B of E, if $A \cup B = E$ and $A \cap B = \phi$, then $B = E - A$.

proof: First we prove that $B \subset E - A$. If $x \in B$, then the second hypothesis of the theorem yields $x \notin A$. Since $x \in B \subset E$, we have $x \in E$; thus $x \in E$ and $x \notin A$, which means that $x \in E - A$. Consequently, $B \subset E - A$.

To show that $E - A \subset B$, let $x \in E - A$. Thus $x \in E$ and $x \in A$. But since $x \in E$ and $x \notin A$ and $E = A \cup B$ it follows that $x \in B$. Hence $E - A \subset B$ and the theorem is proved.

q.e.d.

corollary: For each subset A of E, $E - (E - A) = A$.

proof: Exercise.

An important theorem concerning complements is

Theorem 9. (De Morgan)   For all subsets A and B of a set E,

  1. $E - (A \cup B) = (E - A) \cap (E - B)$, and

  2. $E - (A \cap B) = (E - A) \cup (E - B)$.

EXERCISE

Illustrate the meaning of the theorem with Venn diagrams.

Before proving the theorem, it will be instructive to examine the expressions $x \notin A \cup B$ and $x \notin A \cap B$.   Let us return to the example A = set of positive integers, B = set of integers less than eleven.   Since A and B between them include all the integers, if $x \notin A \cup B$ then x is not an integer.   Consequently, x can be neither an element of A nor an element of B.

For any sets A, B, if $x \notin A \cup B$, then $x \notin A$ and $x \notin B$.   For, if x were an element of A, then by definition of $A \cup B$, $x \in A \cup B$, contrary to hypothesis.   Similarly, if $x \in B$, then $x \in A \cup B$, again a contradiction.

Again using the above example, let us examine the expression $x \notin A \cap B$. $A \cap B$ is the set of integers $\{1, 2, \ldots, 10\}$ and $x \notin A \cap B$ means $x \neq 1$, $x \neq 2$, $\ldots$, $x \neq 10$.   Consequently, either x is not an integer at all, or if it is, then x is not one of the integers $1, 2, \ldots, 10$. In the latter case, x is either an integer greater than ten or $\leq$ zero. Thus, the assertion $x \notin A \cap B$ leads to the following alternatives:

    (i) x is an integer > 10, or

    (ii) x is an integer ≤ 0, or

    (iii) x is not an integer.

If (i) holds, $x \notin B$; if (ii) holds, $x \notin A$; if (iii) holds, $x \notin A \cup B$, since $A \cup B$ is the entire set of integers.   These three alternatives together comprise the meaning of

$$x \notin A \text{ or } x \notin B.$$

Thus from $x \notin A \cap B$ we deduce $x \notin A$ or $x \notin B$.   Similar reasoning shows that for arbitrary sets A, B, if $x \notin A \cap B$, then $x \notin A$ or $x \notin B$.

EXERCISE

Prove the last statement.

proof of theorem 9:   1. We wish to prove that

$$E - (A \cup B) \subset (E - A) \cap (E - B)$$

and

$$(E - A) \cap (E - B) \subset E - (A \cup B).$$

To prove the first of these inclusions let $x \in E - (A \cup B)$; then $x \in E$ and $x \notin A \cup B$. By the foregoing discussion $x \notin A \cup B$ yields $x \notin A$ and $x \notin B$. Thus, from $x \in E - (A \cup B)$ we deduce

$$x \in E \text{ and } x \notin A \text{ and } x \notin B,$$

or

$$(x \in E \text{ and } x \notin A) \text{ and } (x \in E \text{ and } x \notin B).$$

But $x \in E$ and $x \notin A$ means $x \in E - A$, and similarly $x \in E$, $x \notin B$ means $x \in E - B$. Thus, for all x such that $x \in E - (A \cup B)$ we have $x \in E - A$ and $x \in E - B$, whence (Definition 10) $x \in (E - A)$ $\cap (E - B)$. Therefore

$$E - (A \cup B) \subset (E - A) \cap (E - B).$$

To prove the second inclusion, we note that $x \in (E - A) \cap (E - B)$ means $x \in E - A$ and $x \in E - B$, hence $x \in E$ and $x \notin A$, and $x \in E$ and $x \notin B$. Therefore

$$x \in E \text{ and } x \notin A \text{ and } x \notin B.$$

But $x \notin A$ and $x \notin B$ yield $x \notin A \cup B$. Therefore we have

$$x \in E \text{ and } x \notin A \cup B;$$

that is,

$$x \in E - (A \cup B).$$

Thus, for all x such that $x \in (E - A) \cap (E - B)$ we have $x \in E - (A \cup B)$; therefore

$$(E - A) \cap (E - B) \subset E - (A \cup B).$$

This inclusion, together with the one established before, yields 1.

2. The second result is easily obtained from the first. By 1, with $E - A$ in place of A and $E - B$ in place of B, we have

$$E - ((E - A) \cup (E - B)) = (E - (E - A)) \cap ((E - (E - B));$$

since $E - (E - A) = A, E - (E - B) = B,$

$$E - ((E - A) \cup (E - B)) = A \cap B.$$

Hence

$$E - (E - ((E - A) \cup (E - B))) = E - (A \cap B);$$

therefore

$$(E - A) \cup (E - B) = E - (A \cap B),$$

and this is 2.

q.e.d.

EXERCISES

1. Part 2 of Theorem 9 may be proved directly without recourse to part 1 and the relation $E - (E - A) = A$. Carry out the details of the direct proof.

2. Prove: $(A - B) \cup (B - A) = (A \cup B) - (A \cap B).$

3. Prove: $(A - B) \cup (B - C) = (A \cup B) - (C \cap B).$

4. Let A, B be subsets of E. Prove: $A = \phi$ if and only if
$$B = (A \cap (E - B)) \cup ((E - A) \cap B).$$

5. Prove: For all subsets A, B of E:
   (a) If $A \subset E - B$ then $A \cap B = \phi$; conversely, if $A \cap B = \phi$ then $A \subset E - B$.
   (b) If $A \supset E - B$ then $A \cup B = E$; conversely, if $A \cup B = E$ then $A \supset E - B$.

1.10. REMARKS ON NOTATION AND OTHER MATTERS

So far we have mentioned explicitly only one way of denoting a set, namely the symbol

$$\{\ldots\}$$

where the dots are replaced by a listing of the elements in the set. However, in several places, for example in Definitions 10, 11, 12, and others, it was convenient to denote a set of all things having a certain "property."

EXAMPLES

1. In Definition 10′, we defined "$A \cap B$ is the set of all elements x such that $x \in A$ and $x \in B$." In this case, the property in question is that of being an element of A and an element of B.

2. In Definition 12, we defined "$A \cup B$ is the set of all elements x such that $x \in A$ or $x \in B$." Here the property is that of being an element of A or an element of B.

We now introduce a notation to name the set of all things which have a certain property. Without attempting to make the rather vague notion of a property more precise, let us agree that a statement *about* a particular thing is a meaningful sentence in which a name of the thing occurs once or more than once. For instance, consider

1. Bolivia is a country in South America.
2. Canada is a country in South America.
3. Julius Caesar is a country in South America.

Sentence 1 is a statement about Bolivia (and also about countries in South America); sentence 2 is a statement about Canada; sentence 3 is a statement about Julius Caesar. Notice that a statement about a thing need not be true; thus the statement about Canada is false, as is also the statement about Julius Caesar. Another example:

4. Canada is in Europe and some places in Canada are cold.

This statement is about Canada, and it is false.

Now suppose one takes a statement about some thing and everywhere in the statement replaces the name of the thing in question by an otherwise meaningless symbol (the same symbol being used for each occurrence of the name of the thing). The symbol used is ordinarily a letter from some alphabet. Let us do this to the statements 1, 2, 3, and 4 about Bolivia, Canada, Julius Caesar and Canada, respectively, and let us use x as the symbol. We obtain

1′. x is a country in South America.
2′. x is a country in South America.
3′. x is a country in South America.
4′. x is in Europe and some places in x are cold.

The results are examples of what we call "predicates in the symbol x."

A *predicate in the symbol* x is a collection of symbols which results from a statement about some object, by replacing the occurrences of a name of that object in the statement by x.

(Clearly, there is nothing special about the symbol x; we could have used y or z or α or ⊔ instead.)

We shall use the notation "P_x" for a predicate in the symbol x. Now suppose P_x is a predicate in the symbol x; one can replace each occurrence of x in P_x by the name of some object (the same object in each case) and thereby one obtains a statement about that object. Thus if P_x is "x is a country in South America" we can obtain statements about Bolivia, Canada, Julius Caesar, South America by replacing x in P_x in turn by Bolivia, Canada, Julius Caesar, South America, and so on. We get in turn

> Bolivia is a country in South America.
> Canada is a country in South America.
> Julius Caesar is a country in South America.
> South America is a country in South America.

The statements that can be formed in this way from a predicate in a symbol fall into two types: those which are true and those which are false. Now suppose P_x is a predicate in x:

The notation "$\{x \mid P_x\}$" will be used as a name for the set[5] consisting of all those things, and only those things, such that P_x becomes a true statement when the x's in P_x are replaced by names for the things as explained above.

Thus

$$\{x \mid x \text{ is a country in South America}\}$$

is a set among whose elements are Bolivia, Peru, Chile, e.g.,

$$\text{Peru} \in \{x \mid x \text{ is a country in South America}\}.$$

---

[5] The assumption that such a set exists for every predicate, P_x, can lead to difficulties. In fact, the formal, axiomatic treatments of set theory have been developed for the purpose of avoiding the contradictions which arise from the naive assumption that for each P_x there is a set. However, the predicates formulated in this text will lead to no difficulties.

But

> Canada $\notin \{x \mid x$ is a country in South America$\}$.
> Julius Caesar $\notin \{x \mid x$ is a country in South America$\}$.
> South America $\notin \{x \mid x$ is a country in South America$\}$.

Other examples:

> Norway $\in \{x \mid x$ is in Europe and places in x are cold$\}$.
> Canada $\notin \{x \mid x$ is in Europe and places in x are cold$\}$.

The notation "$\{x \mid P_x\}$" is read "the set of all x such that P_x"; e.g., $\{x \mid x$ is a country in South America$\}$ is read "the set of all x such that x is a country in South America."

We emphasize that any symbol which is otherwise without special meaning in a particular discussion can be used for the symbol of a predicate, e.g., $\{z \mid z$ is a country in South America$\}$ is the same set as $\{x \mid x$ is a country in South America$\}$.

With this new notion, we can rephrase Definitions 10′, 11, 12 and 13, respectively, in the following simple ways:

Definition 10″. $A \cap B = \{x \mid x \in A \text{ and } x \in B\}$.

Definition 11′. $A \cap B \cap C = \{x \mid x \in A \cap B \text{ and } x \in C\}$.

Definition 12′. $A \cup B = \{x \mid x \in A \text{ or } x \in B\}$.

Definition 13′. $A - B = \{x \mid x \in A \text{ and } x \notin B\}$.

EXERCISES

In the following, use the notation "$\{x \mid P_x\}$" to denote the sets described.

1. The set of all natural numbers (to be defined in Chapter 2), which are factors of 12.   What other notation can you give for this set?

2. The set of all natural numbers n such that if $m \in n$ then m is a natural number.   (Just what a natural number is does not affect the problem.)

3. Can you give another name (introduced earlier) for $\{x \mid x \neq x\}$?

1.11. SOME SPECIAL SETS

We shall have occasion to refer to sets which have only a single element. For this purpose, we shall use the notation "$\{x\}$"; thus $\{x\}$ is to be understood as the set containing the element x, and no others. More precisely,

Definition 14. $\{x\} = \{u \mid u = x\}$. The reader should observe how concisely this definition expresses the idea discussed above.

Similarly, $\{x,y\}$ is the set consisting of the elements x and y, i.e.,

Definition 15. $\{x,y\} = \{u \mid u = x \text{ or } u = y\}$.

In a similar fashion, we may define $\{x,y,z\}$. For example,

{Alice Zilch, Sam Snork, Mickey Mouse}

is the set consisting exactly of the three individuals, Alice Zilch, Sam Snork and Mickey Mouse.

Now it may very well happen that a given object or individual has different names. Thus, to a fond mother, a lad may be named "Pierpont," but to his more critical circle of friends, he may be known as "Stinky." What does the symbol

{Pierpont, Stinky}

denote? By our agreed-upon notation, this symbol represents the set containing the individuals named "Pierpont" or named "Stinky." But "Pierpont" and "Stinky" are different names for the same individual; consequently

{Pierpont, Stinky} = {Pierpont} = {Stinky}.

More formally, if $x = y$ (i.e., if x and y are the same object) then

$$\{x,y\} = \{x\} = \{y\}.$$

Similarly, if $x = z$ and $v = w = y$, then

$$\{x,y,z,v,w\} = \{x,v\} = \{x,w\} = \{z,y\} = \text{etc.}$$

EXERCISES

1. Verify the foregoing equations.

2. $\{x\} \cup \{y\} = \{x,y\}$.

3. $\{x,y\} \cup \{z\} = \{x,y,z\}$.

4. Give an example to show that the statement $\{x,y\} \cap \{y,z\} = \{y\}$ may be false in some cases.

5. $\{x,y,z\} = \{x,z,y\} = \{z,x,y\} = $ etc.

6. If $\{x\} = \{y\}$ then $x = y$.

7. If $\{x\} = \{y,z\}$ then $x = y = z$.

8. Suppose A, B, C, D are sets such that $\{A,B\} = \{C,D\}$.   Prove: $A \cap B = C \cap D$ and $A \cup B = C \cup D$.   (Hint: consider the two cases, $A = B$ and $A \neq B$.)

The Set $\{\{x\}\}$.   We have mentioned before that the elements of a set may themselves be sets.   For example, let us denote by "*A*", "*B*", "*C*", "*D*", "*E*" the families living on a particular street in some given city. Then

(1.10) $\{A,B,C,D,E\}$

is a set whose elements are the families in question; each of these families is in turn a set, namely, a set of persons.   The set (1.10) should not be confused with the set

(1.11) $A \cup B \cup C \cup D \cup E$

which is defined by

$$\{x \mid x \in A \text{ or } x \in B \text{ or } \dots \text{ or } x \in E\}.$$

The elements of the set (1.11) are *persons*, whereas the elements of (1.10) are *families*, and therefore *the elements of the set (1.11) are not elements of the set (1.10)*.   On the other hand, each element of (1.10) considered as a set is a subset of (1.11).

Suppose now we select the street, Grandview Boulevard, in the town of Foosland, and it turns out that the sole family residing on that street is the Zilch family.   Then the set of all families living on Grandview Boulevard is

{the Zilch family},

where "the Zilch family" itself denotes a set consisting of persons. Suppose, further, that the sole living member of the Zilch family is one Zenobia Zilch.   Then the words, "the Zilch family," would denote a set consisting of a single element, the Zilch family = {Zenobia Zilch}

and the original set {the Zilch family} which consisted of all the families living on Grandview Boulevard in Foosland could be denoted by

(1.12) $\{\{\text{Zenobia Zilch}\}\}$.

Thus (1.12) represents a set having a single element; that single element, in turn, is a set containing a single element. We must distinguish between Zenobia Zilch as an individual person, {Zenobia Zilch} as the Zilch family, and finally $\{\{\text{Zenobia Zilch}\}\}$ as the set of all families residing on Grandview Boulevard in Foosland.

We may sum up the foregoing discussion by remarking that the symbol

$$\{\{x\}\}$$

denotes the set containing the single element $\{x\}$ and that $\{x\}$ itself is a set containing the single element x.

The Set p(E).   We introduce this discussion by means of an example.   Let E be the set $\{a,b,c\}$ where a, b and c are distinct.   The complete list of subsets of E is

(1.13) $\phi,\ \{a\},\ \{b\},\ \{c\},\ \{a,b\},\ \{a,c\},\ \{b,c\},\ \{a,b,c\}.$

The *set of all subsets of E*, denoted by "$p(E)$," is the set

(1.14) $\{\phi,\ \{a\},\ \{b\},\ \{c\},\ \{a,b\},\ \{a,c\},\ \{b,c\},\ \{a,b,c\}\}.$

Thus each set in the list (1.13) is an element of the set $p(E)$ defined in (1.14).   In a similar fashion, for any set E, we shall denote by "$p(E)$" the *set of all subsets of E*.   We make a formal definition:

Definition 16. $p(E) = \{X \mid X \subset E\}$; $p(E)$ is the *power set of E*.

EXERCISES

1.   Let $F = \{a,b,c,d\}$ and find $p(F)$.

2.   If a set E contains n elements, how many elements are there in $p(E)$?

Now let F be the set defined in Exercise 1 above; then $p(F)$ contains 16 elements.   Let S be a subset of $p(F)$.   Thus, the elements of S are elements of $p(F)$; hence *they are subsets of* F. To make the example more concrete, we know that

$$p(F) = \{\phi, \{a\}, \{b\}, \ldots, \{a,b,c,d\}\}.$$

Then, for instance, S might be the set

(1.15) $S = \{\{a\}, \{a,b\}, \{a,b,c\}\}.$

Clearly, $S \subset p(F)$ and the elements of S are certain of the subsets of F (of course, S could be the set $p(F)$ itself). As a preliminary definition, we put

(1.16) $\bigcup S = \{a\} \cup \{a,b\} \cup \{a,b,c\},$

and evidently

$$\bigcup S = \{a,b,c\}.$$

Note that $\bigcup S$ *is a subset of* F. We now reword (1.16) so as to get it into a more convenient form. Obviously, (1.16) has its limitations because it requires that we be able to list explicitly all the elements of S, and in many cases in which this concept is to be used, it is not possible to do so.

Clearly,

(1.17) $\bigcup S = \{x \mid x \in \{a\} \text{ or } x \in \{a,b\} \text{ or } x \in \{a,b,c\}\}.$

If we take into account the fact that $\{a\}, \{a,b\}, \{a,b,c\}$ are all elements of S, it is possible to restate (1.17) as follows:

(1.18) $\bigcup S = \{x \mid$ there is an element of S which, in turn, contains x as an element$\}.$

Finally, if A is a variable whose range is the set S, (1.18) can be abbreviated by

(1.19) $\bigcup S = \{x \mid$ there exists $A \in S$ such that $x \in A\}.$

Equation (1.19) does not require that we be able to list all the elements. The general definition is

Definition 17. Let E be a set, S a subset of $p(E)$. Then

$$\bigcup S = \{x \mid \text{there exists } A \in S \text{ such that } x \in A\}.$$

In words, $\bigcup S$ is the set of all elements contained in one or more sets A such that $A \in S$. An alternative (and useful) notation to "$\bigcup S$" is "$\bigcup_{A \in S} A$."

EXAMPLE

Let E be the set of all triangles t in the plane of analytic geometry, i.e.,

$$E = \{t \mid t \text{ is a triangle in the plane of analytic geometry}\},$$

and define S by

$$S = \{A \mid A \in p(E) \text{ and every triangle } t \in A \text{ has a vertex at the origin}\}.$$

By its definition, $S \subset p(E)$. Then

$$\cup S = \{t \mid \text{there exists an } A \in S \text{ such that } t \in A\};$$

$\cup S \subset E$, and in fact

$$\cup S = \{t \mid t \text{ is a triangle in the plane of analytic geometry and } t \text{ has a vertex at the origin}\}.$$

EXERCISE

Prove the last statement.

In a similar fashion, we can give a general definition of intersection. Again, for purposes of motivation, we use $F = \{a,b,c,d\}$ and define S by (1.15). As a preliminary definition let

$$\cap S = \{a\} \cap \{a,b\} \cap \{a,b,c\};$$

then $\cap S = \{a\} \subset F$. Clearly, $\cap S$ is also given by

$$\cap S = \{x \mid x \in \{a\} \text{ and } x \in \{a,b\} \text{ and } x \in \{a,b,c\}\},$$

or more simply by

$$\cap S = \{x \mid x \in A \text{ for all } A \in S\}.$$

This last equation does not require that we be able to list all the elements of S.

Definition 18. Let E be a set, S a subset of $p(E)$. Then

$$\cap S = \{x \mid x \in A \text{ for all } A \in S\}.$$

We also denote $\cap S$ by " $\underset{A \in S}{\cap} A$."

EXERCISES

1. What is the set $\cap\, S$ for the example following Definition 17?

2. With $E = \{a,b,c\}$ define $S \subset p(E)$ in such a way that $\cup\, S \neq E$ and $\cap\, S \neq \phi$. Define $T \subset p(E)$ so that $\cup\, T = E$ and $\cap\, T = \phi$.

3. Let $E = \{a,b,c,d\}$, and define S so that $\cup\, S = E$, $\cap\, S \neq \phi$.

4. Prove: For all sets E, $\cup\, p(E) = E$ and $\cap\, p(E) = \phi$.

5. Let E be any set, $S = \{A,B\}$ where $A, B \in p(E)$. What can you say about $\cup\, S$? About $\cap\, S$?

6. Let S be a set of sets. Prove:
(a) $\displaystyle\bigcap_{B\in S} (A \cup B) = A \cup (\bigcap_{B\in S} B)$;

(b) $\displaystyle\bigcup_{B\in S} (A \cap B) = A \cap (\bigcup_{B\in S} B)$.

7. Let S and T be sets of sets satisfying the condition: for each $A \in S$ there is a $B \in T$ such that $A \subset B$. Prove that $\displaystyle\bigcup_{A\in S} A \subset \bigcup_{B\in T} B$.

1.12.   ORDERED PAIRS

Consider the justly popular game of sand-lot baseball. In this informal version of the national pastime, the positions are frequently taken by the players in a haphazard way. The pitcher in one inning or game is likely to be the catcher in another, and vice versa. Thus, to state that the two-element set {Jones, Smith} = {Smith, Jones} is the battery for a given team yields no information as to who is pitcher and who is catcher. There is an essential difference between the statement

(1.20)   {Jones, Smith} (= {Smith, Jones}) is the battery

and the statement

(1.21)   Jones and Smith comprise the battery and Smith is
the pitcher, Jones is the catcher.

Indeed, (1.21) informs us not only that (a) the battery is a two-element set, but also that (b) the two elements play different parts.

(1.20) and (1.21) illustrate the difference between the *unordered pair*, i.e., a set consisting of two elements, and the *ordered pair*, a set consisting of two elements whose roles are quite distinct. Of course, one could describe an ordered pair as a two-element set in which the two distinct roles of the elements are distinguished verbally. But this description offends our purist hearts! We should like to do everything, including ordered pairs, in terms of the basic concepts of *set* and *element of a set*.

Before defining the term "ordered pair" we require a theorem concerning the set $\{\{x\}, \{x,y\}\}$.

Theorem 10. If $\{\{x\}, \{x,y\}\} = \{\{z\}, \{z,w\}\}$ then $x = z$ and $y = w$. Conversely, if $x = z$ and $y = w$, then $\{\{x\}, \{x,y\}\} = \{\{z\}, \{z,w\}\}$.

proof: The second statement is easy. For, if $x = z$ and $y = w$, then $\{x\} = \{z\}$ and $\{x,y\} = \{z,w\}$; hence $\{\{x\}, \{x,y\}\} = \{\{z\}, \{z,w\}\}$.

To prove the first statement, note that

(1.22) $\{x\} = \{x\} \cap \{x,y\}$ and $\{x,y\} = \{x\} \cup \{x,y\}$.

Now suppose $\{\{x\}, \{x,y\}\} = \{\{z\}, \{z,w\}\}$. By Exercise 8, page 39, and by (1.22)

(1.23) $\{x\} = \{x\} \cap \{x,y\} = \{z\} \cap \{z,w\} = \{z\}$

and

$$\{x,y\} = \{x\} \cup \{x,y\} = \{z\} \cup \{z,w\} = \{z,w\}.$$

By Equations (1.23), $x = z$.

case 1: $y \neq z$. Then, since $y \in \{z,w\}$, it follows that $y = w$. Therefore, $x = z$ and $y = w$.

case 2: $y = z$. Since $x = z$, we have $x = y$; therefore $\{x\} = \{x,y\} = \{z,w\}$. Hence, $x = y = z = w$.

q.e.d.

Definition 19. The set $\{\{x\}, \{x,y\}\}$ is the *ordered pair* x,y and is denoted by "(x,y)." Thus,

$$(x,y) = \{\{x\}, \{x,y\}\}.$$

The element x is the *initial component*, the element y is the *final component* of the ordered pair (x,y).

The precise meaning to be attached to Definition 19 is as follows:

Whenever the notation "(symbol, symbol)" appears, it denotes a certain set, and the symbol to the left of the comma between the parentheses denotes the object which is the initial component of the ordered pair, and the symbol to the right of the comma between the parentheses denotes the object which is the final component of the ordered pair.[6]

As a matter of convenience, because "left" is a shorter word than "initial," and "right" is a shorter word than "final" (at least in number of syllables), one often says that "x is the left component of (x,y)" and that "y is the right component of (x,y)." This is all right *if one bears clearly in mind that the "left" and "right" refer only to the relative positions of x and y on the printed page and not to the relative position of the things denoted by "x" and "y" in space.* Indeed, the things denoted by "x" and "y" might not even be objects about which it is sensible to speak of location, like ϕ or $\{\phi, \{\phi\}\}$. (The latter is a set which in the next chapter will be given the shorter name "2".)

Returning to the example of a baseball-team battery, we see it is possible to pair the members of the battery so that we know which is the pitcher, which the catcher. If we agree that ordered pairs for batteries will be denoted by

$$\{ \{\text{pitcher}\}, \{\text{pitcher, catcher}\} \},$$

then it is clear from

$$(\text{Smith, Jones}) = \{ \{\text{Smith}\}, \{\text{Smith, Jones}\} \}$$

that Smith is the pitcher and Jones is the catcher.

EXERCISES

1. Prove: If $\{ \{x\}, \{x,y\} \} = \{ \{x\} \}$ then $x = y$, and conversely.

2. Prove the assertion made in footnote 6 that if $x \neq y$ then $(x,y) \neq (y,x)$.

3. Give examples of ordered pairs occurring in elementary mathematics.

[6] Notice that if (x,y) is an ordered pair, neither x nor y are *elements* of (x,y), but x and y are *components* of (x,y). The elements of (x,y) are $\{x\}$ and $\{x,y\}$. The distinction must be carefully drawn. If x and y are distinct objects, then there are two distinct ordered pairs whose components are x and y, namely (x,y) and (y,x), but there is only one set whose elements are precisely x and y, namely $\{x,y\}$, which is the same set as $\{y,x\}$.

1.13. CARTESIAN PRODUCTS, RELATIONS

Definition 20. Let A and B be sets. The *Cartesian product*, $A \times B$, of A and B, is the set of all ordered pairs (x,y) such that $x \in A$, $y \in B$. Thus

$$A \times B = \{(x,y) \mid x \in A \text{ and } y \in B\}.$$

EXAMPLES

1. Suppose four couples are gathered for an evening's dancing, and the rule is established that each man is paired with each woman, in turn, for exactly one dance. If M is the set of men, W the set of women, then $M \times W$ is the set of dancing partnerships. (How many dances take place during the evening?)

2. A familiar example of a Cartesian product taken from elementary mathematics is the set of points of the coordinate plane. Each point of the coordinate plane is an ordered pair (x,y) of real numbers where x is the abscissa, y the ordinate of the point. Thus the coordinate plane may be defined as the Cartesian product $A \times B$ where A and B are both the set of all real numbers.

EXERCISES

1. Let A, B, C, D be sets. Describe the following sets: $(A \times A) \times A$, $A \times (A \times A)$, $(A \times A) \times (A \times A)$, $(A \times B) \times C$, $(A \times C) \times B$, $A \times (B \times C)$, $(A \times B) \times (C \times D)$, $((A \times B) \times C) \times D$. Give specific examples of each of the above products.

2. Prove that for any set A, $A \times \phi = \phi$.

Now, let M be the set of all living men and W the set of all living women. Then $M \times W$ is the set of all living couples, each consisting of exactly one man and exactly one woman. For certain of these pairs (x,y), $x \in M$, $y \in W$, it will be true that the man, x, is married to the woman, y, and for other pairs this will not be true. Let A be the set of all ordered pairs $(x,y) \in M \times W$ such that x is married to y. Then we see that the relation "is married to" has been used to define a certain subset A of $M \times W$. (Note: *relation* has not yet been defined.)

Conversely, the designation of a subset of a Cartesian product $U \times V$ can be used to obtain a relation between U and V.   As an example, let U and V both be the set of children in the Zilch family; Alicia (age 10), Bessie (age 7), Caroline (age 4), Drusilla (age 1).   The Cartesian product $U \times V$ consists of the ordered pairs (using initials to denote the children),

$$(A,A),\ (A,B),\ (A,C),\ (A,D),\ (B,A),\ (B,B),\ (B,C),\ (B,D),$$
$$(C,A),\ (C,B),\ (C,C),\ (C,D),\ (D,A),\ (D,B),\ (D,C),\ (D,D).$$

Now let W be the subset of $U \times V$ defined by

$$W = \{(A,B),\ (A,C),\ (A,D),\ (B,C),\ (B,D),\ (C,D)\}.$$

If we look at the meanings that the letters have, then it is clear that the subset W can be regarded as meaning the relation "is older than." (It may also be possible to regard W in different ways.   See Exercise 1, page 48.)   Thus, the selection of the foregoing subset is tantamount to defining a relation between U and V.

Definition 21.   A *relation C between sets A and B* is a subset of $A \times B$; $C \subset A \times B$.   In particular, if $B = A$, then C is called a "relation on A."   The *domain of the relation C* is the set $\{x \mid (x,y) \in C\}$, and the *range of the relation C* is the set $\{y \mid (x,y) \in C\}$.   We use the notation "$\mathfrak{D}(C)$" to denote the domain of C and "$\mathfrak{R}(C)$" to denote the range of C. Thus

$$\mathfrak{D}(C) = \{x \mid (x,y) \in C\},$$
$$\mathfrak{R}(C) = \{y \mid (x,y) \in C\}.$$

From the definitions of "domain," "range," and "relation," it is clear that

$$\mathfrak{D}(C) \subset A \text{ and } \mathfrak{R}(C) \subset B.$$

The foregoing illustrations have been intended to motivate our definition of *relation*.   However, it may not yet be clear just why we have made our definition in this apparently artificial way.   Our reasons are as follows:

First of all, it was our intention to erect all of the mathematics in this book upon the theory of sets as a base.   We could have introduced the concept of a relation in some other way; but then, in view of our objective, we would have been obligated to show that the concept of relation so introduced could be formulated in terms of set theory.   By our definition, we have eliminated the need for such a proof.

Second, suppose C and C' are two given relations between sets A and B. An important question is "Are the relations C and C' the same?" In terms of our definition, this question is simply "Are the two subsets, C, C' of $A \times B$ the same?" If we had chosen some other definition of the word "relation" we might have considerable difficulty in proving that two given relations either are or are not the same. However, since the relations C and C' are sets, to prove that $C = C'$, we need only show, according to the Axiom of Extensionality, that $C \subset C'$ and $C' \subset C$.

If C is a relation between A and B, the symbol "xCy" will be interpreted as meaning "$(x,y) \in C$"; it is customary to read "xCy" as "x is in the relation C to y." If $(x,y) \notin C$, we write

$$x \mathrel{\not\!C} y.$$

EXERCISES

1. Consider the set of Zilch offspring described in the illustration preceding Definition 21. Suppose that if child X is older than child Y, then X runs faster than Y. Suppose, further, that if X runs faster than Y, then X is older than Y. Describe the relation "runs faster than" as a subset of a Cartesian product. What can you say concerning the relations "is older than" and "runs faster than" in this illustration?

2. Describe sets A, B and C so that the relation C corresponds to "is a son of"; what are $\mathfrak{D}(C)$ and $\mathfrak{R}(C)$?

3. Do the same for "is greater than," "is less than."

Let us return briefly to the example $M \times W$, where M is the set of all living men, W the set of all living women, $A \subset M \times W$ is the relation of all ordered pairs $(x,y) \in M \times W$ such that x is married to y. Then y is married to x. Therefore, along with the relation A, we may consider a set B of ordered pairs (y,x) where $(y,x) \in W \times M$ and $(y,x) \in B$ means that the woman y is married to the man x. More briefly, we can define $B \subset W \times M$ by

$$(y,x) \in B \text{ means } (x,y) \in A.$$

The set B is the *inverse relation* of A. Our formal definition is

Definition 22.   Let R be a relation between sets A and B.   The *inverse relation* of R is the relation $S \subset B \times A$ where $S = \{(y,x)|(x,y) \in R\}$. We shall usually denote the inverse relation of R by "R^{-1}."

EXERCISE

Prove: If T is a relation between A and B then $\mathfrak{D}(T^{-1}) = \mathfrak{R}(T)$ and $\mathfrak{R}(T^{-1}) = \mathfrak{D}(T)$.

The relations of greatest importance are *functions* (or *mappings*), *binary operations* and *equivalence relations*.

1.14.   FUNCTIONS (OR MAPPINGS)

The definition of function frequently given in elementary texts approximates the following:

(1.24)   If there is a rule which associates with each value of a variable x in a range of values, one and only one value of a variable y, then y is called a single-valued function of x.   One writes

$$y = f(x),$$

for each x.

Although this version of the function-concept is a good heuristic starting point, it suffers the deficiency of vagueness (what is a "rule which associates"?).   Also, as interpreted in elementary texts, it leads to the belief that every function is given by a formula, and this concept is inadequate for many purposes.

We can avoid difficulty and ambiguity by seizing upon the essential idea of (1.24) and reformulating it as a set-theoretic concept.   This idea is:

For each element x in a certain set, there is one and only one element y in some set.

Thus, the function concept has to do with a set of ordered pairs (x,y), such that each x is the left member of only one ordered pair in the set. With the foregoing as a guide, we state the definition of function.

Definition 23. *A subset f of $A \times B$ such that*

(i) *for each $x \in A$ there is a $y \in B$ such that $(x,y) \in f$,*
(ii) *for each $x \in A$ and for each y and $z \in B$, if $(x,y) \in f$ and $(x,z) \in f$ then $y = z$,*

is a single-valued function of (or, from) A INTO B or a mapping of A INTO B.

remarks:

1. The only functions considered in this book are single-valued.   We shall abbreviate "single-valued function" simply to "function."

2. The terms "function" and "mapping" are synonymous and are used interchangeably.   A common notation for a mapping f of A into B is

$$f : A \longrightarrow B.$$

3. Since functions f and g are sets, we can deduce $f = g$ from $f \subset g$ and $g \subset f$, by the Axiom of Extensionality.

4. A function is a relation, as is easily seen from a comparison of Definitions 21 and 23.   If $f : A \longrightarrow B$, then the domain of f, $\mathfrak{D}(f) = A$.   Note that the range of f, $\mathfrak{R}(f)$, is a *subset* of B.

EXAMPLES AND EXERCISES

1. $A = \{1,2,3,4\}$, $B = \{1,2,3\}$, $f = \{(1,1), (2,1), (3,2), (4,2)\}$; $\mathfrak{R}(f) = \{1,2\} \subset B$.

2. A and B as in Example 1, $f = \{(1,1), (2,1), (3,1), (4,1)\}$; $\mathfrak{R}(f) = \{1\} \subset B$.

3. $C = \{1,2,3,4\}$, $D = \{1\}$, $g = \{(1,1), (2,1), (3,1), (4,1)\}$; in this case not only do we have $\mathfrak{R}(g) \subset D$ but also $\mathfrak{R}(g) = D$.   Notice that for the function f of the second example and the function g of the third, $f = g$ although $B \neq D$.   However, $\mathfrak{R}(f) = \mathfrak{R}(g)$.

4. $A = $ set of all real numbers $ = B$, $f = \{(x,x^2) \mid x \in A\}$. $\mathfrak{R}(f)$ is the set of all nonnegative real numbers; that is, the set of all real numbers greater than or equal to zero; $\mathfrak{R}(f) \subset B$.

5. E = set of all real numbers, F = set of all nonnegative real numbers, $h = \{(x,x^2) \mid x \in E\}$; $\Re(h) = F$.

6. H = set of all nonnegative real numbers = J,

$$k = \{(x,x^2) \mid x \in H\}.$$

For the functions f and h of Examples 4 and 5 we have $f = h$. But $f = h \neq k$. (Why?)

7. $A = B$ = set of all real numbers, $f = \{(x, \sin x) \mid x \in A\}$. In this case, $\Re(f)$ is the set of all real numbers between -1 and $+1$ inclusive.

8. A = the set of all real numbers between -1 and $+1$ inclusive = B, $F = \{(x, \sqrt{1 - x^2}) \mid x \in A\} \cup \{(x, -\sqrt{1 - x^2}) \mid x \in A\}$. What is $\Re(F)$? Is F a function from A into B? If the points whose coordinates are the ordered pairs in F are plotted in the coordinate plane, what is the graph?

9. Let A, B be as in Example 8 and let $f = \{(x, \sqrt{1 - x^2}) \mid x \in A\}$. Is f a function from A into B? What is the graph of f?

10. Let A, B be as in Examples 8 and 9, and let

$$g = \{(x, -\sqrt{1 - x^2}) \mid x \in A\}.$$

Is g a function from A into B?

11. The set $G = \{(1,2), (1,3), (2,4), (5,7)\}$ with $A = \mathfrak{D}(G) = \{1,2,5\}$ and $B = \{2,3,4,7\}$ is not a function from A into B. Why? Is $G - \{(1,2)\}$ a mapping from A into B?

12. Let $A = B$ be any set and define $I = \{(x,x) \mid x \in A\}$. This set is the *identity mapping* or *identity function* on A.

13. Let $S = \{1,2,3\}$ and let $A = S \times S$,

$$A = \{(1,1), (1,2), (1,3), (2,1), (2,2), (2,3), (3,1), (3,2),$$
$$(3,3)\},$$

and let $B = \{1,2,3,4,5,6\}$. Further, let

$$f = \{((1,1),2), ((1,2),3), ((1,3),4), ((2,1),3), ((2,2),4,$$
$$((2,3),5), ((3,1),4), ((3,2),5), ((3,3),6)\}.$$

Is f a mapping from A into B? Does it remind you of any elementary arithmetic process? Can you describe f in a simpler way?

14. Let $A = \{1,2,3\}$ and let $C = \{1,2,3,4,5,6,7,8,9\}$,

$$g = \{((1,1),1), ((1,2),2), ((1,3),3), ((2,1),2), ((2,2),4),$$
$$((2,3),6), ((3,1),3), ((3,2),6), ((3,3),9)\}.$$

Is g a mapping from $A \times A$ into C? Use an elementary arithmetic process to describe g in a simpler way.

15. Let Z be the set of all integers, $Z = \{\ldots, -2, -1, 0, 1, 2, \ldots\}$, and let $A = Z \times Z$; thus $A = \{(a,b) \mid a,b \in Z\}$. Is the set defined by

$$h = \{((a,b),c) \mid (a,b) \in A \text{ and } c = a + b\}$$

a mapping of A into Z?

16. Let A and Z be as in Example 15 and let $k = \{((a,b),c) \mid (a,b) \in A$ and $c = a \cdot b\}$. Is k a mapping of A into Z?

17. Let A be the set of all real numbers, $B = \{0,1\}$, $f = \{(x,0) \mid x$ is a rational number$\} \cup \{(x,1) \mid x$ is an irrational number$\}$. Clearly, f is a mapping of A into B.

18. $A = $ set of real numbers $= B$,

$$h = \{(x,x) \mid x \geq 0\} \cup \{(x,-x) \mid x < 0\}.$$

What is the common name for this function?

As remarked before, one frequently carries away from elementary mathematics the belief that a function must be expressible in a simple way by means of a single equation, as, for example, $f(x) = \sin x$, for all real x. Examples 1–18 should serve to correct this impression. Indeed, our freedom to define functions is enormous; the only restrictions placed upon our definitions are conditions (i) and (ii) of Definition 23.

Suppose $f : A \longrightarrow B$ where $\Re(f) = B$; this kind of situation occurs several times in the foregoing illustrations. For such mappings we introduce the following special definition:

Definition 23'. Let f be a mapping of A into B such that

$$\Re(f) = B.$$

Then f is a *mapping* (or *function*) of A *ONTO* B.

From Definition 23' it follows at once that every mapping onto is also a mapping into. On the other hand, it is not true that every mapping into is a mapping onto.

EXERCISE

Among Examples 1–18 above, pick out the onto mappings.

A common use for the name of a function is the following: Let f be a function consisting of a certain set of ordered pairs. If a given left member x of an ordered pair in f has y as the right component, then we write "$f(x)$" in place of "y". Thus, for each x, "$f(x)$" is an alternative name for the right component of the ordered pair containing x as the left component. $f(x)$ is the *value* of the function f at x. For example, suppose

$$f = \{(1,2), (3,5), (7,11), (8,-4), (9,2)\}.$$

Then we write

$$f(1) = 2, \quad f(3) = 5, \quad f(7) = 11, \quad f(8) = -4, \quad f(9) = 2,$$

and "2" is a simpler name for the number denoted by "$f(1)$;" "5" is a simpler name for the number denoted by "$f(3)$;" etc. With this notation, it is possible to describe the function defined in Example 4, page 50, in the customary way, as

$$f(x) = x^2, \text{ for all real } x.$$

A function f may, in accordance with the new use to which we have applied f, be represented as

$$f = \{(x,f(x)) \mid x \in \mathfrak{D}(f)\}.$$

Let $f : A \longrightarrow B$, and let $(x,y) \in f$ where $x \in A$ and $y \in B$; then y (i.e., $f(x)$) is *the image of* x. On the other hand, if $y \in \mathfrak{R}(f)$, then any element $x \in A$ such that $(x,y) \in f$, i.e., such that $f(x) = y$, is *AN inverse image of* y. If C is a subset of A, then *the image of* C, denoted by "$f(C)$," is the set

$$f(C) = \{y \mid (x,y) \in f \text{ and } x \in C\}.$$

Further, if D is a subset of B, then *the inverse image of* D, denoted by "$f^{-1}(D)$," is the set

$$f^{-1}(D) = \{x \mid (x,y) \in f \text{ and } y \in D\}.$$

EXAMPLES

1. Let A be the set of all real numbers, B the set of all real numbers ≥ 0, $f = \{(x,x^2) \mid x \in A\}$. The image of 0 is 0; the image of 2

is 4; the image of -2 is 4, etc. An inverse image of 4 is 2, and -2 is also an inverse image of 4. If $C = \{x \mid -1 \leq x \leq 1\}$, the image of C is the set $f(C) = \{y \mid 0 \leq y \leq 1\}$. If D is the set of all real numbers greater than or equal to one, then $D \subset B$ and $f^{-1}(D) = \{x \mid x \geq 1\} \cup \{x \mid x \leq -1\}$.

2. Let $f: A \longrightarrow B$. Then the image of A, $f(A) = \Re(f)$; the inverse image of $\Re(f)$ is A.

3. Let A be a set, $g = \{(x,x) \mid x \in A\}$; thus g is the identity function on A. Then, for each $x \in A$, x is its own image and its own inverse image.

EXERCISES

1. Write the function $g = \{(x, \ln x) \mid x$ is a real number and $x > 0\}$ in the customary functional notation. Do the same for the function

$$h = \{(x, 2x^2 + 3x - 1) \mid x \text{ is a real number}\}.$$

2. Let the functions f and g be defined by

$$f(x) = \frac{x^2 - 1}{x - 1}, \ x \text{ real}, \ x \neq 1,$$

$$g(x) = x + 1, \ x \text{ real},$$

respectively. Is $f = g$?

3. Let $h = \left\{\left(x, \dfrac{x^2 - 1}{x - 1}\right) \mid x \text{ is real and } x \neq 1\right\} \cup \{(1,2)\}$ and let $g = \{(x, x + 1) \mid x \text{ real}\}$. Is $h = g$?

4. Let the functions k, m be defined by

$$k(x) = \frac{x^2 - \frac{1}{4}}{x - \frac{1}{2}}, \text{ for all integers } x,$$

$$m(x) = x + \tfrac{1}{2}, \text{ for all integers } x,$$

respectively. Is $m = k$?

5. In the example $h = \{(x, 2x^2 + 3x - 1) \mid x \text{ is real}\}$, what are the inverse images of -1? of 2? of each $y \in \Re(h)$?

6. In the example $g = \{(x, \ln x) \mid x \text{ is real and } > 0\}$, let C be the set of all real numbers ≥ 1.   What is the image, $g(C)$, of C?

7. Let f, g be functions with the same domain A.   Prove: If $f(x) = g(x)$ for all $x \in A$ then $f = g$; conversely, if $f = g$, then $f(x) = g(x)$ for all $x \in A$.

8. Suppose f is a function and $x \in \mathfrak{D}(f)$.   Is there any difference between $f(\{x\})$ and $f(x)$?   If $y \in \mathfrak{R}(f)$ what is $f^{-1}(\{y\})$?

9. Let $f: A \longrightarrow B$ where $\phi \notin A$.   What is $f(\phi)$?   What is $f^{-1}(\phi)$?

10. If f is the function $f = \{(x,x^2) \mid x \text{ real}\}$, and $S = \{x \mid -1 \leq x\}$, $T = \{x \mid x \leq 1\}$, what is $f(S \cup T)$? $f(S \cap T)$? $f(S) \cup f(T)$? $f(S) \cap f(T)$?

On occasion, it is useful to create from a given mapping a new one, called a "restriction" of the original mapping.   To illustrate the idea, consider the sets $A = \{1,2,3,4,5\}$, $B = \{7,9,11,12\}$ and $f: A \longrightarrow B$ defined by $f = \{(1,7), (2,9), (3,9), (4,11), (5,11)\}$.   Further, let $C = \{1,3,4\}$; clearly C is a subset of A.   It is plausible that the mapping f of A into B also yields a mapping of C into B.   If we discard from f those ordered pairs whose left members are not elements of C, we obtain a new set, namely,

$$\{(1,7), (3,9), (4,11)\},$$

and this set (as the reader can verify easily) is a mapping from C into B. The mapping of C into B obtained in this way is the *restriction of f to C*. More generally,

Definition 24. Let $f: A \longrightarrow B$, and let C be a subset of A.   The *restriction of f to C*, denoted by "$f \mid C$," is the set

$$f \mid C = \{(x,y) \mid (x,y) \in f \text{ and } x \in C\}.$$

If f and g are mappings such that g is a restriction of f, then f is an *extension* of g.

EXERCISES

1. In the illustration preceding Definition 24, let $D = \{2,4,5\}$, $E = \{2,3,4\}$, $G = \{1\}$.   What are $f \mid D$, $f \mid E$, $f \mid G$?

2. Let $A = B =$ set of all real numbers, $C =$ set of all integers. Let $f = \{(x,x^2) \mid x \text{ real}\}$. What is $f \mid C$?

3. Let $g = \left\{(x, \tan x) \mid x \text{ real}, x \neq \dfrac{2n + 1}{2} \pi, n \text{ an integer}\right\}$. Let C be the set of all real numbers x such that $-\dfrac{\pi}{2} < x < \dfrac{\pi}{2}$. What is $g \mid C$?

4. Let $\varphi: A \longrightarrow B$ and let C be a subset of A. Prove that $\varphi \mid C$ is a mapping.

5. With the same hypotheses as in Exercise 4, prove that

$$\varphi \mid C = \varphi - \{(x,y) \mid (x,y) \in \varphi \text{ and } x \notin C\}.$$

To introduce the next concept, let us take an example from elementary mathematics. Suppose f is the function defined for all real numbers x by $f(x) = \sin x$; suppose g is the function defined for all real numbers x by $g(x) = x^2$; and finally, suppose that h is defined for all real numbers x by $h(x) = (\sin x)^2$. Thus

$$f = \{(x, \sin x) \mid x \text{ is a real number}\},$$
$$g = \{(x,x^2) \mid x \text{ is a real number}\},$$
$$h = \{(x,(\sin x)^2) \mid x \text{ is a real number}\}.$$

Let us calculate $g(f(x))$ for every value of x. We have $g(f(x)) = g(\sin x) = (\sin x)^2 = h(x)$, so for all real numbers x, $g(f(x)) = h(x)$; in other words, the value of h at each element of its domain is obtained by first applying f to that element and then applying g to the result. The reader has probably guessed that to construct this example we used f and g to compute h so that the result $h(x) = g(f(x))$, for all x, would hold. The function h obtained from g and f is called the "composite of g and f." We now give the formal definition in terms of ordered pairs.

Definition 25. Let $f: A \longrightarrow B$ and $g: B \longrightarrow C$. Then the *composite of g and f* is the set

$$\{(x,g(f(x))) \mid x \in A\}.$$

The notation for the composite of g and f is "$g \circ f$" so that our definition may be written

(1.25) $$g \circ f = \{(x,g(f(x))) \mid x \in A\}.$$

We have remarked before that if $h:E \longrightarrow F$ and if $(x,y) \in h$, then for each x it is customary to write "$h(x)$" in place of "y", so that

$$h = \{(x,h(x)) \mid x \in E\}.$$

With this convention, $g \circ f$ becomes

$$g \circ f = \{(x,(g \circ f)(x)) \mid x \in A\},$$

hence by (1.25)

(1.26) $(g \circ f)(x) = g(f(x))$, for all $x \in A$.

Equation (1.26) is an alternate way of defining the composite of g and f.
    Note that if $f:A \longrightarrow B$ and $g:C \longrightarrow D$ where $B \neq C$, then the composite of g and f is not defined.

EXAMPLE

Let $f = \{(3,2), (4,3), (5,7)\}$, $g = \{(2,1), (3,9), (7,14)\}$ where $A = \{3,4,5\}$, $B = \{2,3,7\}$, $C = \{1,9,14\}$.   Then

$$(g \circ f)(3) = g(f(3)) = g(2) = 1,$$
$$(g \circ f)(4) = g(f(4)) = g(3) = 9,$$
$$(g \circ f)(5) = g(f(5)) = g(7) = 14.$$

Thus, $g \circ f = \{(3,1), (4,9), (5,14)\}$ with $\mathfrak{D}(g \circ f) = \{3,4,5\}$ and $\mathfrak{R}(g \circ f) = \{1,9,14\}$.

EXERCISES

1. If $f = \{(1,1), (2,1), (3,2), (4,2), (5,2)\}$ and $g = \{(1,1), (2,1), (3,1)\}$ where $A = B = C = \{1,2,3,4,5\}$, is $g \circ f$ defined?   Is there a restriction, g^*, of g such that $g^* \circ f$ is defined?

2. Let $f:A \longrightarrow B$ and $g:B \longrightarrow C$.   Prove that $g \circ f$ is a mapping of A into C.

3. If f and g (Exercise 2) are both onto, prove that $g \circ f$ is also onto.

An important result concerning the composite of functions is

Theorem 11. Let $f: A \longrightarrow B$, $g: B \longrightarrow C$, $h: C \longrightarrow D$. Then

$$h \circ (g \circ f) = (h \circ g) \circ f.$$

In short: *composition of mappings is associative.*

proof: By Exercise 2 above, we know that

$$g \circ f \text{ is a mapping of } A \text{ into } C,$$

hence

$$h \circ (g \circ f) \text{ is a mapping of } A \text{ into } D.$$

Similarly,

$$h \circ g \text{ is a mapping of } B \text{ into } D,$$

hence

$$(h \circ g) \circ f \text{ is a mapping of } A \text{ into } D.$$

We must prove that the sets $h \circ (g \circ f)$ and $(h \circ g) \circ f$ are equal. Using Exercise 7, page 55, we need only prove that

$$[(h \circ g) \circ f] (x) = [h \circ (g \circ f)] (x), \text{ for all } x \in A.$$

Now, for each $x \in A$

$$[h \circ (g \circ f)] (x) = h((g \circ f) (x)). \qquad \text{(Equation 1.26)}$$

But $(g \circ f) (x) = g(f(x))$ so that

$$[h \circ (g \circ f)] (x) = h(g(f(x))).$$

Since $f(x) \in B$, we have

$$h(g(f(x))) = (h \circ g)(f(x)),$$

and since $h \circ g$ is a mapping of B into D

$$(h \circ g)(f(x)) = [(h \circ g) \circ f] (x).$$

From this sequence of equations, it follows that

$$[(h \circ g) \circ f] (x) = [h \circ (g \circ f)] (x), \text{ for all } x \in A;$$

therefore, $(h \circ g) \circ f = h \circ (g \circ f)$.

q.e.d.

Among mappings, those which are termed "one-one correspondences" are fundamental. Essentially, the one-one correspondence is the device which we use in counting. As such, it is probably one of the oldest intellectual devices known to man; it has been used by exceedingly primitive peoples as well as by contemporary persons living in highly developed societies.   Counting on one's fingers is one of the uses to which it is put, and, at the same time, it is an indispensable tool of sophisticated mathematics.

It is well known that many primitive peoples do not have names for numbers beyond, say, three or five.   Suppose a primitive herdsman has 27 goats; how does he count the number of animals in his herd, and determine loss or perhaps theft, if he has number names only up to and including five?   He does so by admitting his goats to a corral one at a time, and for each goat entering the enclosure, he drops exactly one pebble into a gourd.   Thus the number of goats in the flock is precisely represented by the number of pebbles in the gourd.   Let F be the set of goats in the herdsman's flock, P the set of all pebbles in the gourd. We define the subset h of $F \times P$ by

$(g,p) \in h$ means p is the pebble dropped into the gourd when the goat g entered the corral.

Clearly, h is a function onto. For if (g,p) and $(g,p') \in h$ then $p = p'$; therefore condition (ii) of Definition 23 is satisfied.   Since for each goat $g \in F$ there is a pebble p in the gourd, condition (i) is also satisfied. Finally, since the herdsman tosses exactly one pebble into the gourd for each goat entering the corral, it follows that $\Re(h) = P$, hence h is onto. In addition,

$$\text{if } (g,p),\ (g',p) \in h \text{ then } g = g'.$$

In words, *the same pebble is not dropped into the gourd for different goats!* It is a property such as the last which distinguishes one-one correspondences among mappings.

Definition 26.   *A one-one correspondence between sets A and B is a mapping f of A onto B such that*

$$\text{if } (x,y) \text{ and } (z,y) \in f \text{ then } x = z.$$

Thus, if $f : A \longrightarrow B$ and is a one-one correspondence between A and B, then f is a relation between A and B (i.e., a subset of $A \times B$) such that

   (i) if $(x,y),\ (x,z) \in f$ then $y = z$;
   (ii) if $(x,y),\ (z,y) \in f$ then $x = z$;
   (iii) $\mathfrak{D}(f) = A$ and $\Re(f) = B$.

The expressions "one-one correspondence," "one-one mapping," and "one-one function" are synonymous.

EXAMPLES

1. If A, B and f are respectively $\{1,2,3,4,5\}$, $\{2,4,6,8,10\}$ and $\{(1,2),$ $(2,4),$ $(3,6),$ $(4,8),$ $(5,10)\}$ then $f:A \longrightarrow B$ and is a one-one correspondence between A and B.

2. Let M be the set of all living, married men in the United States, W the set of all living, married women in the United States. Define the subset g of $M \times W$ by

$$g = \{(x,y) \mid (x,y) \in M \times W \text{ and } x \text{ is married to } y\}.$$

Then the mapping $g:M \longrightarrow W$ is a one-one correspondence between M and W. (It is essential to exclude bigamists.)

Definition 26′. Let A, B, C be sets such that $B \subset C$, and let f be a one-one correspondence between A and B. Then f is also a *one-one correspondence of A INTO C* (also, one-one mapping of A INTO C or one-one function of A INTO C).

EXERCISES

1. Give some examples of one-one mappings into, but not onto.

2. Let f, g be one-one correspondences, $f:A \longrightarrow B$ and $g:B \longrightarrow C$. Prove that $g \circ f$ is a one-one correspondence. If both f, g are onto, so is $g \circ f$; if g is into but not onto, so is $g \circ f$.

3. Let A, B be sets. Prove that there is a one-one correspondence between $A \times B$ and $B \times A$. Is it always true that $A \times B = B \times A$? Is it ever true?

4. Let A, B, C be sets. Prove that there is a one-one correspondence between $(A \times B) \times C$ and $A \times (B \times C)$.

5. Prove that for any set A there is a one-one correspondence between $A \times \{\phi\}$ and A.

6. If φ is a one-one correspondence between A and B, and if $x \notin A$, $y \notin B$, then $\psi = \varphi \cup \{(x,y)\}$ is a one-one correspondence between $A \cup \{x\}$ and $B \cup \{y\}$ such that $\psi(x) = y$.

7. If φ is a one-one correspondence between A and B and if $x \in A$, then $\varphi - \{(x, \varphi(x))\}$ is a one-one correspondence between $A - \{x\}$ and $B - \{\varphi(x)\}$.

8. Prove: If φ is a one-one correspondence between A and B, if $x \in A$ and $y \in B$, then there exists a one-one correspondence ψ between A and B such that $\psi(x) = y$. (Hint: Use Exercises 6 and 7.)

In Definition 22, we defined the inverse of a relation. Since every function is a relation, this definition yields also a definition of the inverse of a function. For our purposes, we shall not be concerned with the inverses of arbitrary mappings. However, the inverses of one-one correspondences are of great importance and will be used frequently.

EXERCISES

1. What is the inverse of the one-one correspondence described in the "goats and pebbles" example (page 59)?

2. Prove that the inverse of a one-one correspondence between sets A and B is a one-one correspondence between the sets B and A.

3. Let f be a one-one correspondence between A and B and let g be a one-one correspondence between B and C. Then (Exercise 2, page 60) we know that $g \circ f$ is a one-one correspondence between A and C. Hence, $(g \circ f)^{-1}$ is a one-one correspondence between C and A. Prove that

$$(g \circ f)^{-1} = f^{-1} \circ g^{-1}.$$

Another mapping that concerns us is the *binary operation*. As with one-one correspondences, the reader has used instances of such operations, possibly without being aware that they are mappings. To illustrate the idea, let us consider a very elementary example.

For each integer a, and for each integer b, there is one and only one integer c such that

$$a + b = c.$$

Another way to put the same statement is this:

With every ordered pair, (a,b), of integers, is associated one and only one integer c, the sum of a and b, denoted by "$a + b$".

In discussing mappings, we saw that associations of this kind were made precise by the concept of mapping. It therefore seems plausible that certain operations, as, for example, addition, could be described as mappings.

Definition 27. Let A be a set. A mapping φ of $A \times A$ into A is a *binary operation on A*.

EXAMPLES AND EXERCISES

1. Let Z be the set of all integers. Which of the sets

$$k = \{((a,b), a + b) \mid (a,b) \in Z \times Z\},$$
$$l = \{((a,b), a \cdot b) \mid (a,b) \in Z \times Z\},$$
$$m = \{((a,b),c) \mid (a,b) \in Z \times Z, b \neq 0, c = a \div b\},$$
$$n = \{((a,b), a + b - a \cdot b) \mid (a,b) \in Z \times Z\},$$

is a binary operation on Z?

2. Let $A = \{0,1,2,3\}$. Are the following sets binary operations on A?

(i) $f = \{((a,b),0) \mid (a,b) \in A \times A\}$;
(ii) $g = \{((a,b),1) \mid (a,b) \in A \times A\}$;
(iii) $h = \{((a,b),a) \mid (a,b) \in A \times A\}$.

3. Let Θ be the set of all odd integers. Is

$$\varphi = \{((a,b), a + b) \mid (a,b) \in \Theta \times \Theta\}$$

a binary operation on Θ? Is $\psi = \{((a,b), a \cdot b) \mid (a,b) \in \Theta \times \Theta\}$ a binary operation on Θ?

4. Give examples of binary operations on the set of real numbers; the set of complex numbers.

At this point, the reader may have the feeling that he is being sold a bill of goods on binary operations. After all, addition is addition, multiplication is multiplication, and why all the fuss? (By the way, what are "ordinary" addition and multiplication? We shall have much to say about these topics in Chapters 2, 3, 4 and 5.) The reason is that in mathematics there are binary operations of importance other than the familiar operations of elementary arithmetic. A number of these more esoteric binary operations will be studied in considerable detail—some of them are important in the study of physics and technological applications as well as in mathematics. Also, as we have emphasized, it was our intention to describe elementary mathematics entirely in terms of set theory. Definition 27 defines one of the concepts that enables us to carry out this program.

1.15.   EQUIVALENCE RELATIONS AND PARTITIONS

The final concepts to be studied in this chapter are *equivalence relations* and *partitions;* these are closely connected to each other.

Definition 28.   An *equivalence relation on a set A* is a relation E on A such that

(i) for all $x \in A$, $(x,x) \in E$;
(ii) for all $x, y \in A$, if $(x,y) \in E$ then $(y,x) \in E$;
(iii) for all $x, y, z \in A$, if $(x,y) \in E$ and $(y,z) \in E$ then $(x,z) \in E$.

Using the notation introduced in connection with relations, we can restate the above as:

(i′) for all $x \in A$, $x \, E \, x$;
(ii′) for all $x, y \in A$, if $x \, E \, y$ then $y \, E \, x$;
(iii′) for all $x, y, z \in A$, if $x \, E \, y$ and $y \, E \, z$ then $x \, E \, z$.

In practice, it is customary to use "$\sim$" (wiggle) as a symbol for an equivalence relation.[7] With this symbol, we restate Definition 28 in the form in which it ordinarily appears:

Definition 28′.   An *equivalence relation on a set A* is a relation $\sim$ on A such that

(i″) for all $x \in A$, $x \sim x$;
(ii″) for all $x, y \in A$, if $x \sim y$ then $y \sim x$;
(iii″) for all $x, y, z \in A$, if $x \sim y$ and $y \sim z$ then $x \sim z$.

EXAMPLES

1.   Let T be the set of all triangles in the plane of elementary geometry. $T \times T$ is the set of all ordered pairs (x,y) of triangles x, y.

---

[7] It is a little uncomfortable at first to accept "$\sim$" as a symbol for a set. It's something like the lowly anchovy; you must become accustomed to it in order to like it.

Define the equivalence relation $\sim$ on T by: $x \sim y$ if and only if x is similar to y. Since every triangle is similar to itself, $x \sim x$ for all triangles x. If x is similar to y, then y is similar to x; hence, if $x \sim y$, then $y \sim x$. Finally, if x is similar to y and y is similar to z, then x is similar to z. Thus, from $x \sim y$ and $y \sim z$, follows $x \sim z$.

2. Let A be the set of all automobiles, and define the relation $\sim$ by: $x \sim y$ if and only if the automobile x has the same amount of chrome (by area) as automobile y. Verify that $\sim$ is an equivalence relation.

3. Choose a pair of perpendicular lines in the coordinate plane as x- and y-axes and let π be the set $\{P,Q,R, \ldots\}$ of points in the plane. Define the relation $\sim$ on π by: "$P \sim Q$" means "P is on the same line parallel to the y-axis as Q." We consider the y-axis to be parallel to itself. Again, it is easy to verify that $\sim$ is an equivalence relation.

4. Let Z be the set of all integers $\{\ldots, -3, -2, -1, 0, 1, 2, 3, \ldots\}$. From elementary arithmetic, we know that if a, $b > 0$ are two integers, there are unique integers q (quotient) and r (remainder) such that $a = qb + r$ where $0 \leq r < b$. Define the relation $\sim$ on Z by: "$x \sim y$" means "x has the same remainder when divided by 5 as y." For example, $13 \sim 8$, since $13 = 2 \cdot 5 + 3$ and $8 = 1 \cdot 5 + 3$, and therefore the remainder in each case is 3. The relation $\sim$ is an equivalence relation. (Verify!)

EXERCISES

1. As in Example 4, let Z be the set of integers. Define the relation $\sim'$ by: for all $x, y \in Z$, "$x \sim' y$" means "$x - y$ is divisible by 5." Show that $\sim = \sim'$.

2. With Z as above, define $\sim_n$ by: for all $x, y \in E$, "$x \sim_n y$" means "x has the same remainder, when divided by n, as y." Show that $\sim_n$ is an equivalence relation. Define $\sim'_n$ as in Exercise 1 and show that for each integer n, $\sim_n = \sim'_n$.

3. Show that the inverse relation of an equivalence relation is the same relation.

4. Let E be a given set. For all A, $B \in p(E)$, define the subset $R \subset p(E) \times p(E)$ by:

"$(A,B) \in R$" means "there is a one-one correspondence between A and B."

Prove that R is an equivalence relation on $p(E)$.

Equivalence relations have a useful property which we shall illustrate in terms of Examples 3 and 4. In Example 3, for each point $P \in \pi$, let S_P be the subset of all points $Q \in \pi$ such that $Q \sim P$; more briefly

$$S_P = \{Q \mid Q \sim P\}.$$

For a given point P, S_P is easily determined; it consists of those and only those points on the line through P, parallel to the y-axis. If $P \neq Q$, there are two alternatives: either $S_P = S_Q$ or $S_P \cap S_Q = \phi$. The former case arises if P and Q are on the same line parallel to the y-axis; the latter, if P and Q are on different lines parallel to the y-axis. Since for all points P, $P \sim P$, therefore $P \in S_P$, it follows that each point P in π is contained in at least one subset $S_P \subset \pi$. Thus, beginning with the equivalence relation $\sim$ of Example 3, we are able to separate π into subsets S_P, S_Q, S_R, ... such that

(a) for all P, $Q \in \pi$ exactly one of the equations $S_P = S_Q$ or $S_P \cap S_Q = \phi$ holds;

(b) for each $P \in \pi$ there is a subset, $S_P \subset \pi$, containing P as an element.

Briefly, (b) can be stated

$$\pi = \bigcup_{P \in \pi} S_P.$$

Observe that property (a) for the subsets S_P, S_Q, ... is a property which is not enjoyed by *every* collection of subsets of a given set. For, if A, B are subsets of a set C, then there are three mutually exclusive alternatives: either $A = B$, or $A \cap B = \phi$, or $A \neq B$ and $A \cap B \neq \phi$ (give examples). For the subsets S_P, S_Q, ... satisfying (a), the third alternative is excluded.

Conversely, suppose a collection of subsets of π is given such that (a) and (b) hold. We shall illustrate by means of an example that such a collection of subsets enables us to define an equivalence relation on π. Consider the subsets of π defined by

$$S_{0,0} = \{P:(x,y) \mid 0 \le x < 1, 0 \le y < 1\},$$
$$S_{1,0} = \{P:(x,y) \mid 1 \le x < 2, 0 \le y < 1\},$$
$$S_{0,1} = \{P:(x,y) \mid 0 \le x < 1, 1 \le y < 2\},$$
$$\cdots\cdots\cdots\cdots\cdots\cdots\cdots\cdots\cdots\cdots\cdots\cdots$$
$$S_{i,j} = \{P:(x,y) \mid i \le x < i + 1, j \le y < j + 1 \mid\},$$

for all integers i, j. $(P:(x,y)$ is the point whose coordinates are (x,y).) Each $S_{i,j}$ consists of all the points of a square including the lower and the left-hand boundaries. The vertices of the square $S_{i,j}$ are the points with coordinates (i,j), $(i + 1, j)$, $(i + 1, j + 1)$, $(i, j + 1)$. Clearly, two such squares are either identical or else they have no points in common. Hence, property (a) holds. Moreover, for any point P with coordinates (x,y), there is always a pair of integers i, j such that $i \le x < i + 1, j \le y < j + 1$. Consequently, every point is contained in a subset $S_{i,j}$. Thus, property (b) holds.

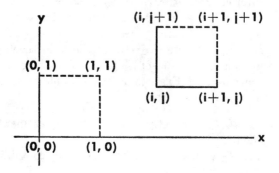

Now, using the subsets $S_{i,j}$, $(i, j = 0, \pm1, \pm2, \ldots)$ we shall define an equivalence relation $\sim'$ on P. For all points P, Q define $P \sim' Q$ by: "$P \sim' Q$" means "P and Q are in the same subset $S_{i,j}$." With the notation of Definition 21, this definition of $\sim'$ can be given in the form: For all points P, Q, "$(P,Q) \in \sim'$" means "P and Q are in the same subset $S_{i,j}$." For each point P, P is in the same subset $S_{i,j}$ as P; hence $P \sim' P$. If P is in the same subset $S_{i,j}$ as Q, then Q is in the same subset as P, so that (ii'') holds. Finally, it is clear that (iii'') holds. Consequently, $\sim'$ is an equivalence relation.

EXERCISE

In the foregoing discussion, equivalence relations $\sim$ and $\sim'$ have been defined in the plane π. Prove that $\sim \ne \sim'$.

If we had chosen as subsets of π having properties (a), (b), those subsets $S_P, S_Q, \ldots$ which arose from the original equivalence relation $\sim$, then we could prove that the new relation, $\sim'$, defined in terms of subsets $S_P, S_Q, \ldots$ is the same as the old one, $\sim$.

Turning to Example 4 (see page 64), for each integer x we define

$$Z_x = \{y \mid y \text{ is an integer and } y \sim x\}.$$

Thus, Z_x is the subset of Z consisting of all integers y which have the same remainder as x when divided by 5. Now the possible remainders, when x is divided by 5, are $0, 1, 2, 3, 4$. Therefore, if x has the remainder 0, then Z_x consists of all the multiples of 5; if x has the remainder 1, then Z_x consists of all the integers $5n + 1$, and similarly if x has a remainder 2, 3, or 4. Consequently, the subsets which we obtain in this way are

$$Z_0 = \{5n \mid n \in Z\},$$
$$Z_1 = \{5n + 1 \mid n \in Z\},$$
$$Z_2 = \{5n + 2 \mid n \in Z\},$$
$$Z_3 = \{5n + 3 \mid n \in Z\},$$
$$Z_4 = \{5n + 4 \mid n \in Z\},$$

and there are no others; for each integer x, Z_x is one of the subsets $Z_0, \ldots, Z_4$. Moreover, no two of the subsets $Z_0, \ldots, Z_4$ have any elements in common. Thus, using the relation $\sim$, we have separated Z into subsets Z_x having the following properties:

(a') for any two integers x, y, either $Z_x = Z_y$ or else $Z_x \cap Z_y = \phi$;
(b') $Z = \bigcup\limits_{x \in Z} Z_x$.

These properties are completely analogous to the properties (a), (b) for the subsets of P in the preceding example.

Conversely, if Z is separated into a collection of subsets having properties (a') and (b'), then one can show that such a collection of subsets can be used to define an equivalence relation on Z.

EXERCISE

Exhibit a separation of Z into a collection of non-empty subsets (other than the one in the text) satisfying (a') and (b'); use the separation to define an equivalence relation on Z. Show that if the collection of subsets $Z_0, \ldots, Z_4$ is used to define an equivalence relation $\sim'$, then $\sim' = \sim$.

We shall prove, below, a sequence of theorems which show that the situation illustrated for the sets π and Z are by no means restricted to these two sets. First we shall introduce a definition which will simplify our terminology.

Definition 29. A *partition* $\mathcal{P}$ of a set A is a subset of $p(A)$ such that

    (i) if $S \in \mathcal{P}$ then $S \neq \phi$,
    (ii) if $S, T \in \mathcal{P}$ then either $S = T$ or $S \cap T = \phi$,
    (iii) $\bigcup \mathcal{P} = A$.

The sets $Z_0, \ldots, Z_4$ form a partition of Z, and the sets $S_{i,j}$ ($i, j = 0, +1$, $+2, \ldots$) form a partition of π.

Theorem 12(a). Let $\sim$ be an equivalence relation on A. For each $x \in A$ let

$$S_x = \{y \mid y \in A \text{ and } y \sim x\}.$$

Further, let

$$\mathcal{P}(\sim) = \{S \mid S \text{ is one of the sets } S_x \text{ for some } x \in A\}.$$

Then $\mathcal{P}(\sim)$ is a partition of A.

proof: By its definition, $\mathcal{P}(\sim)$ is a subset of $p(A)$. In order to prove that $\mathcal{P}(\sim)$ is a partition of A, we prove that $\mathcal{P}(\sim)$ satisfies conditions (i), (ii) and (iii) of Definition 29.

    (i) Let $S \in \mathcal{P}(\sim)$; by definition of $\mathcal{P}(\sim)$, $S = S_x$ for some $x \in A$. Since $\sim$ is an equivalence relation, it follows that $x \sim x$. Since $S_x = \{y \mid y \in A, \text{ and } y \sim x\}$, we have $x \in S_x = S$. Consequently, $S \neq \phi$ and (i) holds.
    (ii) Let $S, T \in \mathcal{P}(\sim)$. Then there exist $x, y \in A$ such that $S = S_x$, $T = S_y$. We shall prove that if S_x and S_y have any element at all in common, then $S_x = S_y$. This will exclude the alternative that both $S_x \neq S_y$ and $S_x \cap S_y \neq \phi$ hold. Therefore, it must follow that exactly one of the two equalities $S_x = S_y$ or $S_x \cap S_y = \phi$ holds for all S_x and S_y in $\mathcal{P}(\sim)$.
    Suppose S_x and S_y have an element z in common, i.e.,

$$z \in S_x \cap S_y.$$

Since $z \in S_x \cap S_y$, we deduce $z \in S_x$ and $z \in S_y$. Hence, by definition of S_x and S_y, $z \sim x$ and $z \sim y$. But, since $\sim$ is an equivalence relation, from $z \sim x$ we get $x \sim z$. Thus, $x \sim z$ and $z \sim y$; by (iii″) of Definition 28′ it follows that $x \sim y$ and therefore by (ii″), $y \sim x$.

To prove that $S_x = S_y$ we show that $S_x \subset S_y$ and $S_y \subset S_x$. The first inclusion is deduced easily as follows: for all $u \in S_x$, $u \sim x$. But $x \sim y$, therefore by (iii″) for all $u \in S_x$ we have $u \sim y$. Hence, by the definition of S_y, for all $u \in S_x$ we have also $u \in S_y$, whence $S_x \subset S_y$. By reversing the roles of S_x and S_y we can also prove that $S_y \subset S_x$. Hence, $S_x = S_y$ and condition (ii) of Definition 29 is satisfied.

(iii) To prove that $\cup \, \mathcal{P}(\sim) = A$, we show that for each $x \in A$ there is an $S \in \mathcal{P}(\sim)$ such that $x \in S$. As we have seen before, always $x \in S_x$. By definition of $\mathcal{P}(\sim)$, $S_x \in \mathcal{P}(\sim)$ for all x. Hence, $A \subset \cup \, \mathcal{P}(\sim)$. The reverse inclusion is obvious. This concludes the proof of (iii) and therefore of Theorem 12(a).

Theorem 12(b).   Let $\mathcal{P}$ be a partition of A. Define $E(\mathcal{P})$ to be the set of all ordered pairs $(x,y) \in A \times A$ such that

there exists a set $S \in \mathcal{P}$ such that $x \in S$ and $y \in S$.

Then $E(\mathcal{P})$ is an equivalence relation on A.

proof: We show that the three conditions of Definition 28 are satisfied by the subset $E(\mathcal{P})$ of $A \times A$.

(i) Let x be an element of A. Since $\mathcal{P}$ is a partition of A, there exists a set $S \in \mathcal{P}$ such that $x \in S$. Since $x \in S$, it follows that $x \in S$ and $x \in S$. Hence, by definition of $E(\mathcal{P})$, $(x,x) \in E(\mathcal{P})$. Thus, the first condition holds.

(ii) Suppose $(x,y) \in E(\mathcal{P})$. By definition of $E(\mathcal{P})$, there is a set $S \in \mathcal{P}$ such that $x \in S$ and $y \in S$. Hence, for the same set $S \in \mathcal{P}$, $y \in S$ and $x \in S$. By definition of $E(\mathcal{P})$, this means $(y,x) \in E(\mathcal{P})$; i.e., the second condition is satisfied.

(iii) Finally, let (x,y) and $(y,z) \in E(\mathcal{P})$. We show that $(x,z) \in E(\mathcal{P})$. Since $(x,y) \in E(\mathcal{P})$, there is a set $S \in \mathcal{P}$ such that $x, y \in S$; since $(y,z) \in E(\mathcal{P})$, there is a set $T \in \mathcal{P}$ such that $y, z \in T$. Hence, $y \in S$ and $y \in T$, i.e., $y \in S \cap T \neq \phi$. But, by hypothesis, $\mathcal{P}$ is a partition of A. Therefore (condition (ii), Definition 29) $S = T$. Thus there is a set $S \, (= T)$ such that $x, z \in S$. Again, by definition of $E(\mathcal{P})$, $(x,z) \in E(\mathcal{P})$. This concludes the proof of Theorem 12(b).

Theorem 12(c). (α) If $\sim$ is an equivalence relation on A, then

$$E(\mathcal{P}(\sim)) = \sim.$$

(β) If $\mathcal{P}_1$ is a partition of A, then $\mathcal{P}(E(\mathcal{P}_1)) = \mathcal{P}_1$.

Part (α) of the theorem states: suppose we start with a given equivalence relation $\sim$ on A. From this equivalence relation, we construct a partition $\mathcal{P}(\sim)$ of A, in accordance with Theorem 12(a). Then from the partition $\mathcal{P}(\sim)$ we construct an equivalence relation on A in accordance with Theorem 12(b). Then the resulting equivalence relation is the same as the one $\sim$ with which we began the sequence of constructions.

EXERCISE

State in words the meaning of part (β) of the theorem.

proof of theorem 12(c):

(α) To prove that $E(\mathcal{P}(\sim)) = \sim$ we show that $E(\mathcal{P}(\sim)) \subset \sim$ and $\sim \subset E(\mathcal{P}(\sim))$. It will be convenient to carry out the proof as is done in elementary geometry, listing the steps on the left side of the page, the reasons on the right side. We show first that $E(\mathcal{P}(\sim)) \subset \sim$. Let $(u,v) \in E(\mathcal{P}(\sim))$; we want to prove that $(u,v) \in \sim$.

1. Since $(u,v) \in E(\mathcal{P}(\sim))$, there is a set $S \in \mathcal{P}(\sim)$ such that $u, v \in S$.

   1. This is a consequence of the definition of E in Theorem 12(b), since $\mathcal{P}(\sim)$ is a partition (Theorem 12(a)).

2. $S = S_x$, for some x.

   2. Definition of $\mathcal{P}(\sim)$ in Theorem 12(a).

3. $S_x = \{y \mid y \in A \text{ and } y \sim x\}$.

   3. Definition of S_x.

4. Hence, $u \sim x, v \sim x$.

   4. Definition of S_x.

5. Therefore, $u \sim v$, i.e., $(u,v) \in \sim$.

   5. Properties of equivalence relations.

6. Hence, $E(\mathcal{P}(\sim)) \subset \sim$.

   6. Steps 1–5.

Conversely, to prove $\sim\, \subset E(\mathcal{P}(\sim))$, we let $(u,v) \in\, \sim$ and show that $(u,v) \in E(\mathcal{P}(\sim))$.

7. $u \in S_v,\, v \in S_v.$	7. Definition of S_v.
8. Hence, there is a set $S \in \mathcal{P}(\sim)$ containing the elements $u,\, v$.	8. Definition of $\mathcal{P}(\sim)$.
9. $\mathcal{P}(\sim)$ is a partition.	9. Theorem 12(a).
10. $(u,v) \in E(\mathcal{P}(\sim))$.	10. Definition of $E(\mathcal{P}(\sim))$.
11. Therefore $\sim\, \subset E(\mathcal{P}(\sim))$.	11. Steps 7–10.

From 6 and 11 we conclude that $E(\mathcal{P}(\sim)) =\, \sim$.

(β) To prove that $\mathcal{P}(E(\mathcal{P}_1)) = \mathcal{P}_1$, we again use the Axiom of Extensionality.

Let $S \in \mathcal{P}(E(\mathcal{P}_1))$; our object is to show that $S \in \mathcal{P}_1$. By definition of $\mathcal{P}(E(\mathcal{P}_1))$ (see the statement of Theorem 12(a)), $S \in \mathcal{P}(E(\mathcal{P}_1))$ means that $S = \{y \mid y \in A$ and $y\, E(\mathcal{P}_1)\, x$ for some $x \in A\}$. But $y\, E(\mathcal{P}_1)\, x$ means (statement of Theorem 12(b)) that there is a $T \in \mathcal{P}_1$ such that $y \in T$ and $x \in T$. Hence, for each $u \in S$ we have $u \in T$ and therefore $S \subset T$. On the other hand, if $z \notin S$, then $z\, \overline{E(\mathcal{P}_1)}\, x$; therefore there is no element $W \in \mathcal{P}_1$ such that both z and x are elements in W. In particular, $z \notin T$. Thus, from $z \notin S$, we deduce $z \notin T$, whence $T \subset S$. With the inclusion $S \subset T$, established before, it follows that $S = T$. Therefore, $S \in \mathcal{P}_1$ and $\mathcal{P}(E(\mathcal{P}_1)) \subset \mathcal{P}_1$.

We leave it to the reader to complete (β) by showing that $\mathcal{P}_1 \subset \mathcal{P}(E(\mathcal{P}_1))$. By the Axiom of Extensionality, $\mathcal{P}(E(\mathcal{P}_1)) = \mathcal{P}_1$ follows, and Theorem 12(c) is proved.

Theorem 12(d).

(α) If $\sim$ and $\sim'$ are equivalence relations on a set A, then $\sim\, =\, \sim'$ if and only if $\mathcal{P}(\sim) = \mathcal{P}(\sim')$.

(β) If $\mathcal{P}_1$ and $\mathcal{P}_2$ are partitions of A, then $\mathcal{P}_1 = \mathcal{P}_2$ if and only if $E(\mathcal{P}_1) = E(\mathcal{P}_2)$.

proof: (1) If $\sim\, =\, \sim'$, then it follows at once that $\mathcal{P}(\sim) = \mathcal{P}(\sim')$. Similarly, if $\mathcal{P}_1 = \mathcal{P}_2$, then $E(\mathcal{P}_1) = E(\mathcal{P}_2)$.

(2) Next, suppose $\mathcal{P}(\sim) = \mathcal{P}(\sim')$ where $\sim$ and $\sim'$ are equivalence

relations on A. By (1) we have $E(\mathcal{P}(\sim)) = E(\mathcal{P}(\sim'))$. But, by Theorem 12(c), part (α),

$$E(\mathcal{P}(\sim)) = \sim \text{ and } E(\mathcal{P}(\sim')) = \sim'.$$

Hence, $\sim = \sim'$.

(3) Similarly, if $E(\mathcal{P}_1) = E(\mathcal{P}_2)$, where $\mathcal{P}_1$, $\mathcal{P}_2$ are partitions of A, then since $E(\mathcal{P}_1)$ and $E(\mathcal{P}_2)$ are equivalence relations, (1) yields

$$\mathcal{P}(E(\mathcal{P}_1)) = \mathcal{P}(E(\mathcal{P}_2)).$$

By Theorem 12(c), part (β),

$$\mathcal{P}(E(\mathcal{P}_1)) = \mathcal{P}_1 \text{ and } \mathcal{P}(E(\mathcal{P}_2)) = \mathcal{P}_2.$$

Hence, $\mathcal{P}_1 = \mathcal{P}_2$, and Theorem 12(d) is proved.

Theorems 12(a)–(d) comprise Theorem 12.

Definition 30. Let $\mathcal{P}$ be a partition of a set A and let $\sim$ be the equivalence relation on A determined by $\mathcal{P}$ according to Theorem 12(b). An element $S \in \mathcal{P}$ is an *equivalence class* with respect to the relation $\sim$.

1.16. MATHEMATICAL SYSTEMS

The remainder of this book is devoted to a detailed study of several elementary mathematical systems. And, as has been emphasized several times, everything that we do is based upon the theory of sets as developed in this chapter. Thus we shall avail ourselves freely of such concepts as subset, union, intersection, ordered pair, Cartesian product, relation, mapping, and so on.

What do we mean by a mathematical system? We have seen that, starting with a single set, an enormous array of other sets is obtained. For instance, there are all the subsets of the given set; there is the set of all subsets of the given set; there are Cartesian products, unions, and intersections of the foregoing; and these concepts can be applied again and again to define still more sets. Clearly, the array of sets so obtained is huge. Now a mathematical system can be defined roughly as:

> A given set, together with certain sets from the large array of sets that can be formed from the given set, and all of these sets are subject to certain conditions.

The first system that we study in Chapter 2 is the system of *natural numbers*; these are the numbers which the child learns about in the first grade of grammar school, or even earlier.   In subsequent chapters, we shall investigate the systems of the integers, the rational numbers, the real numbers and the complex numbers, respectively.   The second part of this three-part work will be devoted to a study of some of the systems that comprise algebra, and the third part will be concerned with the system, known as elementary (Euclidean) geometry.

2

THE NATURAL NUMBERS

2.1. THE DEFINITION OF THE NATURAL NUMBERS

What is a natural number? From early childhood everyone is familiar with the list: "zero," "one," "two," "three," etc., but what are these words the names of? It is to this question that we propose to

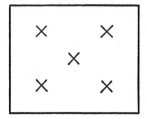
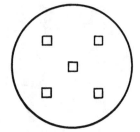

Figure 1

give an answer. Such words as "four," "seventeen," "two hundred eighty-four" are used in English both as adjectives and nouns. Perhaps the adjectival use is the more common. The box in Figure 1 contains *five* crosses, the circle contains *five* squares. If challenged to explain what is meant by this statement, one might say that there exists a one-one correspondence between the crosses in the box and the squares in the circle and that use of the same word "five" in both statements conveys this idea.   In similar fashion, to say of any set A that A contains five elements can be interpreted to mean that there is a one-one correspondence f between A and B, where B is the set of crosses in the box on this page.   This begins to suggest a possible method of defining "five."   We might define the noun "five" to mean a certain set, say the set of crosses in the box in Figure 1, and then define the adjective "five" by saying that a set has five elements if it can be placed in one-one correspondence with the set, five.   In like fashion, we might define "seventeen"; say, seventeen is the set of crosses in Figure 2.

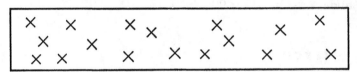

Figure 2

Zero is the set of crosses in Figure 3.   The set of crosses in Figure 3 is the empty set; thus, by the proposed definition, zero is ϕ.

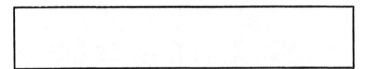

Figure 3

Now there can be some objection to the above procedure.   To define five as a certain standard and fixed set seems to be a good idea, but it would be desirable to find a more suitable (and less perishable) set than the set of crosses in Figure 1.   For one thing, there is more than one copy of this figure: is the standard set, five, Joe Jones' copy or Alice Zilch's?   We might do slightly better by storing a carefully labeled bag of steel balls in the Bureau of Standards, and defining five to be the set of steel balls in that bag.   That would do for five.   Then six, then seven, then eight, . . . , then 2,718,982,394, . . . , then . . . .   No, this is impractical, nay impossible; neither the Bureau of Standards nor the surface of the earth, nor perhaps even the entire universe, can store so many balls.

The definition we proposed for zero, however, seems satisfactory.

* Definition 1. *Zero* is the empty set; i.e., $0 = \phi$.*

(Throughout Chapters 2, 3 and 4, we shall use the device of enclosing the formal statements of the theory within asterisks. Statements not so enclosed are intended as informal and motivational. Thus they supply intuition but are not part of the development of the theory.)

Now, can we find a satisfactory set (with one element) to be the standard *one*? The set $\{0\} = \{\phi\}$ seems satisfactory.

* Definition 2. $1 = \{0\} = 0 \cup \{0\}$.*

EXERCISE

Prove that $1 \neq 0$.

Next, two: $\{0,1\}$ seems suitable; it is certainly a set of imperishable objects. The idea seems to be taking form:

> Definition: $2 = \{0,1\} = 1 \cup \{1\}$
> Definition: $3 = \{0,1,2\} = 2 \cup \{2\}$
> Definition: $4 = \{0,1,2,3\} = 3 \cup \{3\}$
> Definition: $5 = \{0,1,2,3,4\} = 4 \cup \{4\}$.

Each of these definitions is unequivocal, because no notion is used in any of them that has not been previously defined.[1] We could clearly go on in this fashion indefinitely, defining *6* as $5 \cup \{5\}$, *7* as $6 \cup \{6\}$ and so on.

Let us observe that if n is any one of the sets we have defined above, with the exception of 0, then there is a set x such that $n = x \cup \{x\}$. This suggests the following definition:

---

[1] It may be amusing to look at the names of 0, 1, 2, 3, 4, 5 which are formed by using merely ϕ, $\{,\}$ and commas.

$0 = \phi$
$1 = \{\phi\}$
$2 = \{\phi, \{\phi\}\}$
$3 = \{\phi, \{\phi\}, \{\phi, \{\phi\}\}\}$
$4 = \{\phi, \{\phi\}, \{\phi, \{\phi\}\}, \{\phi, \{\phi\}, \{\phi, \{\phi\}\}\}\}$
$5 = \{\phi, \{\phi\}, \{\phi, \{\phi\}\}, \{\phi, \{\phi\}, \{\phi, \{\phi\}\}\}, \{\phi, \{\phi\}, \{\phi, \{\phi\}\}, \{\phi, \{\phi\}, \{\phi, \{\phi\}\}\}\}\}$.

* **Definition 3**.   The set $x \cup \{x\}$ is the *successor* of the set x.   If y is a set and if there is a set x such that y is the successor of x, then y is a *successor*.   For each set x, the successor of x is x'.*[2]

Thus, $1 = 0'$, $2 = 1'$, $3 = 2'$, etc.

EXERCISES

1.   Prove: If $y \in x'$ then $y \in x$ or $y = x$.

2.   Prove: ϕ is not a successor.

We can now turn to the question posed at the beginning of this chapter: "What is a natural number?"   A first answer that might occur to the reader could be: A natural number is a set which is either empty or else can be obtained by applying the successor operation repeatedly to the empty set.   Now this is certainly the right idea but it suffers from several defects, which we shall try to explain.   The weakness lies in the phrase "can be obtained" in this connection.   It seems to suggest that only those sets are natural numbers for which it is possible to write down an explicit definition of the same kind as has been done for 0,1,2,3,4,5.   But if one interprets "can be obtained" in such a way, it becomes very doubtful that $10^{10,000,000}$, say, could be considered a natural number.   We wish to take the position that there is such a natural number as $10^{10,000,000}$.   But *can* the series of definitions which has been started for the small natural numbers be continued until one reaches the definition of $10^{10,000,000}$?   Not in one lifetime, certainly, and probably not in the lifetime of the whole human race.   There is also doubt as to whether the physical universe contains enough material to make the paper to write the definitions on.

Thus the attempt to interpret, carefully, the first proposed definition of natural number leads to difficulties which are quite hard to circumvent.   It might be possible to circumvent these difficulties if one were willing to take a philosophic position which granted some sort of ideal sense to the notion that something can be done even if no one is really capable of doing it.   But, fortunately, there is another alternative.   It is possible to give a definition of natural number which is free of the

[2] Recall that if x is a set, then $\{x\}$ is a set whose sole element is the set x.   Hence $x \cup \{x\}$ is a set whose elements are: (a) all the elements in x and (b) an additional element, namely x itself.

objections we have outlined and which nevertheless is in agreement with our intuitive notions as to which sets should be called "natural numbers."

Before we actually set down our definition of natural number, let us discuss what we should expect of it in order to feel that the definition does agree with our intuitive notions of what it should do. In the first place, the definition will say that a set n is a natural number if it satisfies a certain list of conditions imposed on it. We will then introduce "N" as the notation for the set of all natural numbers. In other words, N is the set whose elements are those sets n which meet the conditions of the definition of natural number. Of course, we could write down any definition we pleased as the definition of natural number, and simply accept as natural number anything that satisfied the definition. But we have a very strong intuitive feeling as to just what sort of thing a natural number should be. Thus, in view of the foregoing discussion, we would be inclined to reject a proposed definition of natural number if, according to the definition, the set of bananas in Peru were to be a natural number, or if, say $2 = \{\phi, \{\phi\}\}$ were not to be a natural number. In other words, we have some fairly clear ideas as to what consequences we expect to deduce from a definition that we are willing to accept as the "right" definition for natural number.

Now, what are some consequences we could expect to deduce from a definition of natural number that is acceptable as the "right" definition?

First, from the definitions of 0,1,2,3,4,5,6,7 and of the term "successor," we should expect that $0 \in N$, hence that $1 = 0' \in N$, hence $2 = 1' \in N$, etc. In short, we expect of the set N of natural numbers that $0 = \phi \in N$; and if $n \in N$ then the successor n' of n is also an element of N.

To facilitate our discussion we introduce

* Definition 4. Let A be such that for all x, if $x \in A$ then the successor, x', of x is also an element of A. Then A is a *hereditary* set.*

The statement preceding Definition 4 is then the assertion that

(α) $0 \in N$ and N is hereditary.

Second, it is possible that there are sets, other than N, which are hereditary and contain 0 as an element. Suppose X is such a set. Then $0 \in X$ and by Definition 4, $1 = 0' \in X$, hence $2 = 1' \in X$, and so on. Thus, it is reasonable to expect of any hereditary set X which contains 0 that it shall contain all the natural numbers. In other words,

(β)   If X is a hereditary set such that $0 \in X$ then $X \supset N$.

There are many other things that one expects from the "right" definition of natural number, but surely (α) and (β) should be among the anticipated consequences. We now deduce an important conclusion from (α) and (β).

* Theorem 1. If M and N are sets, each of which satisfies both conditions (α) and (β), then $M = N$.*

Before proving Theorem 1, let us see what is remarkable about it. Suppose it is possible to give a "right" definition of natural number in two different ways.[3] Let M be the set of all natural numbers according to the first "right" definition, and let N be the set of all natural numbers according to the second "right" definition. Then, by the preceding discussion, both M and N must satisfy conditions (α) and (β). By Theorem 1 it follows that $M = N$. Thus, from Theorem 1, we deduce that any "right" definition will do, so long as (α) and (β) are consequences of that definition.

*proof of theorem 1: It will be convenient to analyze the meaning of the hypotheses of the theorem. The assumption that M satisfies (α) and (β) means

(α_M) M is a hereditary set containing the element 0, and
(β_M) if X is any hereditary set containing the element 0 then $X \supset M$.

The assumption that N satisfies (α) and (β) means

(α_N) $0 \in N$, N is hereditary, and
(β_N) if Y is any hereditary set such that $0 \in Y$ then $Y \supset N$.

Our object is to prove that $M \supset N$ and $N \supset M$.

By (α_M) $0 \in M$ and M is hereditary; hence, by (β_N), $M \supset N$. Similarly, by (α_N), $0 \in N$ and N is hereditary; hence, by (β_M), $N \supset M$. Therefore $M = N$ as asserted, and this completes the proof of Theorem 1.*

Once the definition of the set of natural numbers has been given, the troublesome thought that there might be several sets satisfying our requirements for the natural numbers has been ruled out.

How can we find a satisfactory definition of natural number? We have a clear idea of what a natural number is and we have explicitly defined several sets which are to be instances of natural numbers.

[3] This can, in fact, be done.

Natural numbers are, at any rate, certain sets. Perhaps we can find a list of properties, shared by all our instances (0,1,2,3,4,5,6,7), such that these properties will fully characterize those sets which we wish to call natural numbers.

Consider the set 4, for example; the elements of 4 are 0,1,2,3. Among these elements one observes that $0 \in 1$, $0 \in 2$, $0 \in 3$, $1 \in 2$, $1 \in 3$, $2 \in 3$; this follows from the definitions of 0,1,2,3. We observe that if x and y are elements of 4, then either $x \in y$, or $x = y$, or $y \in x$. Further, these possibilities are mutually exclusive; thus $2 \in 3$, but $2 \neq 3$ and $3 \notin 2$, and likewise $3 = 3$ but $3 \notin 3$. Similar remarks could be applied to 4, 5, 6 and 7. We seem to have found a property that should be shared by all natural numbers. Let us make a definition.

* Definition 5. A set A is $\in$-*ordered* means:

(a) for all x and for all y in A at least one of

$$(\dagger) \quad x \in y, x = y, \text{ or } y \in x$$

holds, and

(b) at most one of ($\dagger$) holds.$_{*}$[4]

EXERCISE

Verify that 0, 1, 2, 3, 4, 5 are $\in$-ordered.

Now let us take the set 6, for example, and consider a nonempty subset of 6, say $A = \{5,2,3\}$. The element $2 \in A$ is a rather special element of A in that 2 is an element of each of the other elements of A, namely $2 \in 3$ and $2 \in 5$. The reader will convince himself, after considering a few examples, that a similar phenomenon occurs for any of the nonempty subsets of the sets 0, 1, 2, 3, 4, 5, 6, 7. This suggests a new definition.

* Definition 6. An element x in B, such that for all $y \in B$, $y \neq x$, one has $x \in y$, is a *leading element* of B.*

---

[4] It is customary to write a definition such as Definition 5 in the following abbreviated form: A set A is $\in$-*ordered* means: for all x and for all y, one and only one of the following holds:

$$x \in y, x = y, \text{ or } y \in x.$$

In other words, "one and only one" is an abbreviation for "at least one . . . and at most one . . . ."

Thus, 2 is a leading element of the set $A = \{5,2,3\}$. And each nonempty subset of 0, 1, 2, 3, 4, 5, 6, 7 has a leading element.

So far we have found two properties which all the sets we would like to call natural numbers seem to share, namely, $\in$-orderedness and the property that every nonempty subset of a natural number has a leading element.

To obtain the next property consider $4 = \{0,1,2,3\}$. Clearly, $2 \in 4$; but, since 2 was defined as $2 = \{0,1\}$ we see that $2 \subset 4$. Similarly, $3 \in 4$ and $3 = \{0,1,2\} \subset 4$. And, in fact, the corresponding phenomenon occurs for each of 0, 1, 2, etc.; namely, if n is such a set and if $x \in n$, then $x \subset n$.

EXERCISES

1. Examine a few of the sets 0, 1, 2, 3, 4, 5, 6, 7 for the above described property.

2. Construct sets S such that for some x, $x \in S$ but $x \subset S$ is false.

Finally, we observe that 1 is the successor of 0, $1 = 0'$, likewise $2 = 1'$, $3 = 2'$, etc. It seems that each nonempty natural number is a successor. Further, it appears that each nonempty element of a natural number is a successor.

By examining several sets which we feel should be called "natural numbers" we have found a list of properties shared by these sets. Conversely, we shall see that the sets having these properties are precisely those sets which we want to regard as natural numbers. (Indeed, it will be proved that the set, whose elements are all the sets having these properties, has properties (α) and (β).) We now adopt the following as our formal definition of natural number:

* Definition 7. A *natural number* is a set n such that

(A) n is $\in$-ordered,
(B)   every nonempty subset of n possesses a leading element,
(C)   if $x \in n$ then $x \subset n$,
(D)   if n is not empty then n is a successor,
(E)   if $x \in n$ and x is not empty then x is a successor.

The set of all natural numbers is denoted by "N."$_*^5$

The definition of natural number just given is such as to satisfy our most scrupulous demands as to precision and unambiguity.

Condition (C) in Definition 7 may be a bit troublesome, but the reader should recall that an example of sets satisfying this condition has been given.

EXERCISES

1. Prove: If n' is a natural number, so is n.

2. Prove: If n is a natural number and if $x \in n$, then $x' \subset n$.

It is perhaps well to remind the reader (again) that only the statements enclosed within asterisks are part of our formal development; all the rest is informal motivational material. We ask the reader to bear in mind that henceforth whenever we say that n is a natural number, we mean that n is an object which satisfies the conditions of the formal definition; and we mean precisely that; no more, no less. We emphasize this point because so much has been said informally concerning the nature of natural numbers. We must, therefore, exercise care never to assume in the formal development that natural numbers possess some property which has only been asserted in the informal part. It might be well for the reader to go back and reread the material enclosed in asterisks, skipping the intervening remarks. He will observe that it is a very small part of the total thus far. It consists of practically nothing but definitions. We shall soon remedy this and start to prove some theorems.

We now turn our attention to proving that N satisfies (α) and (β), (page 78). This will be the substance of Theorems 3 and 5.

* Theorem 2. If n is a natural number, then $n \notin n$.

proof: We give an indirect proof of the theorem. Thus, we assume that n is a natural number such that $n \in n$, and show that this assumption produces a contradiction.

---

[5] For other definitions of "natural number," or more generally, "ordinal number," see: "Transfinite Zahlen," by H. Bachmann, *Ergebnisse der Mathematik und Ihrer Grenzgebiete*, Springer-Verlag; and, "A Definition of Ordinal Numbers," by J. R. Isbell, *Amer. Math. Monthly*, January 1960.

First, recall that for every set S, it is true that $S = S$; in particular, therefore, $n = n$. Since n is a natural number, it is $\in$-ordered (condition (A) of Definition 7). On the other hand, if n satisfies the hypothesis that $n \in n$, we would have $n \in n$ and $n = n$. This contradicts Definition 5. Therefore $n \in n$ must be false, whence $n \notin n$ is true.

q.e.d.

Theorem 3. $\phi \in N$ and N is hereditary.

proof: We demonstrate (a) $\phi \in N$, and (b) if $n \in N$ then

$$n' = n \cup \{n\} \in N.$$

(a) To prove that $\phi \in N$, we show that ϕ satisfies each of conditions (A)–(E) of Definition 7.

(A) ϕ is $\in$-ordered. For, if x, $y \in \phi$, then one and only one of the following holds: $x \in y$, $x = y$, $y \in x$.

(B) We verify that every nonempty subset of ϕ has a leading element. Since there are no nonempty subsets of ϕ, they all have a leading element.

In a similar fashion, each of (C)–(E) is rendered trivial by the fact that ϕ is the empty set.

(b) Suppose $n \in N$, i.e., suppose n is a set for which conditions (A)–(E) are satisfied; we show that n' also satisfies these conditions.

(A) Suppose x, $y \in n'$. Then $x \in n \cup \{n\}$ so that (Exercise 1, page 77) $x \in n$ or $x = n$. Likewise, $y \in n$ or $y = n$. The following are all the alternatives:

(i) $x \in n$ and $y \in n$, (ii) $x \in n$ and $y = n$, (iii) $x = n$ and $y \in n$, (iv) $x = n$ and $y = n$.

(i) $x \in n$ and $y \in n$. Since n satisfies condition (A), it follows that $x \in y$, $x = y$, or $y \in x$, and only one of these possibilities is realized. Therefore, (A) holds.

(ii) If $x \in n$ and $y = n$, then $x \in y$. We show further that $x \neq y$ and $y \notin x$. Suppose $x = y$; then $y \in n$, and since $y = n$ we have $n \in n$, contrary to Theorem 2. Therefore, $x \neq y$. Next, suppose $y \in x$. We remember that n satisfies (C) and, since $x \in n$, it follows that $x \subset n$. But then $y \in x$ gives $y \in n$; since we already have $y = n$, this yields $n \in n$, a contradiction. Therefore $y \notin x$.

(iii) If $x = n$, $y \in n$, reasoning in all respects identical with the above shows that $y \in x$ but $y \neq x$ and $x \notin y$.

(iv) If $x = n$, $y = n$, then $x = y$. We show that $x \notin y$ and $y \notin x$. But since x and y are both equal to n, the statements $x \notin y$ and $y \notin x$ are both merely the true statement $n \notin n$.

In summary, (i)–(iv) show that for all x and $y \in n'$, exactly one of $x \in y$, $x = y$, $y \in x$ holds; therefore condition (A) is satisfied for n'.

(B) Assuming that every nonempty subset of n has a leading element, we prove that every nonempty subset of n' also has a leading element. To this end, let $\phi \neq X \subset n'$ and let $Y = X - \{n\}$. Clearly $Y \subset X$ and $Y \subset n$.

If $Y = \phi$, then $X = \{n\}$ and n is a leading element of X.

Suppose $Y \neq \phi$. Since $Y \subset n$, it follows that Y has a leading element, say z. We claim that z is a leading element of X. First note that since $Y = X - \{n\}$, $X \subset Y \cup \{n\}$. Now let $a \in X$; then $a \in Y$ or $a = n$. If $a \in Y$, then $z \in a$ or $z = a$, since z is a leading element of Y. On the other hand, if $a = n$, then $z \in Y \subset n$, whence $z \in a$. Thus, for all $a \in X$, $z \in a$ or $z = a$, and therefore z is a leading element of X.

Thus, n' satisfies condition (B).

(C) Suppose $x \in n'$. Then, since $n' = n \cup \{n\}$ either $x \in n$ or $x \in \{n\}$ (i.e., $x = n$). If $x \in n$, since n satisfies (C), we have $x \subset n \subset n'$. If $x = n$, since $n \subset n'$, we have $x \subset n'$. Thus, any element of n' is a subset of n' and n' satisfies (C).

(D) Since n' is the successor of n, it is obviously a successor.

(E) Suppose $x \in n'$ and $x \neq \phi$. Then either $x \in n$ or $x = n$. If $x \in n$, then since n satisfies (E), x is a successor. If $x = n$ then n is not empty, and since n satisfies (D), $n = x$ is a successor.

Summing up, n' satisfies (A)–(E); therefore $n' \in N$.

q.e.d._*

Before turning to Theorem 4, we emphasize that, because of Theorem 3, it is now possible to assert that the sets 0, 1, 2, . . . so far defined are natural numbers. We also take the occasion to extend the list of defined natural numbers to 9.

* Definition 8. $6 = 5 \cup \{5\} = 5'$. $7 = 6'$. $8 = 7'$. $9 = 8'$.

Theorem 4. 0, 1, 2, 3, 4, 5, 6, 7, 8 and 9 are natural numbers.

proof: By Theorem 3, $0 = \phi \in N$. Also, since N is hereditary, if $x \in N$, it follows that $x' \in N$ (Definition 4). Then, since $0 \in N$, $1 = 0' \in N$,

$$\text{and since } 1 \in N, 2 = 1' \in N$$
$$\text{and since } 2 \in N, 3 = 2' \in N$$
$$\text{and since } 3 \in N, 4 = 3' \in N$$
$$\text{and since } 4 \in N, 5 = 4' \in N$$
$$\text{and since } 5 \in N, 6 = 5' \in N$$
$$\text{and since } 6 \in N, 7 = 6' \in N$$
$$\text{and since } 7 \in N, 8 = 7' \in N$$
$$\text{and since } 8 \in N, 9 = 8' \in N$$

q.e.d.∗

A remark on notation will be useful at this point. If we so desired, we could go on inventing new symbols and then assign them as names to certain natural numbers in exactly the fashion we have been doing. Thus, we might continue with the following definitions: ⊐ = 9′, ⊔ = ⊐′, ⊠ = ⊔′. Here "⊐," "⊔" and "⊠" are arbitrary symbols which have been assigned definite meanings just as "0," "1," "2," "3," "4," "5," "6," "7," "8," "9" are arbitrary symbols which have been assigned definite meanings. The reader is aware that, in naming the natural numbers beyond 9, it is customary not to introduce new arbitrary symbols such as "⊐," "⊔," "⊠," but to employ an extremely clever procedure, whereby several of the symbols "0," "1," "2," "3," "4," "5," "6," "7," "8," "9," are placed in juxtaposition to form names of natural numbers. Thus this system assigns the name "10" to the natural number which we called "⊐," the name "11" to the natural number ⊔, the name "12" to the natural number ⊠, and the name "4874" to a certain natural number whose definition is too tedious to set down explicitly. This system is known as the decimal system for the natural numbers. There are also other systems for naming natural numbers, such as the binary system, which employs only two symbols rather than ten.

Before turning to Theorem 5, which shows that (β) is a consequence of our definition of the natural numbers, it is useful to have a

∗lemma: If X is a hereditary set, if $\phi \in X$, and if n is a natural number, then $n \subset X$.

proof: Suppose $n \not\subset X$; we shall prove that this supposition leads to a contradiction and, therefore, $n \subset X$. If $n \not\subset X$, there is an element in n which is not an element in X (Definition 6′, Chapter 1). Hence, $n - X \neq \phi$. Set $Y = n - X$; Y is a nonempty subset of n, and therefore, by (B) of Definition 7, Y has a leading element, say y. Then $y \notin X$ and since $\phi \in X$ (hypothesis), it follows that $y \neq \phi$. Thus, y is a nonempty element of n, and hence, by condition (E) of Definition 7, y is the successor of some set, say z; i.e., $y = z \cup \{z\}$. Therefore $z \in y$; and since $y \in n$, we have $y \subset n$, whence $z \in n$. So y and z are both elements of n. Since n is $\in$-ordered, and since $z \in y$, it follows that $y \notin z$ and $y \neq z$. Now, $z \notin Y$. For, if $z \in Y$, then since y was chosen as a leading element of Y and since $y \neq z$, we would have $y \in z$, a contradiction. Further, since $z \notin Y = n - X$ and $z \in n$, it follows that $z \in X$. But by hypothesis, X is hereditary, and therefore $z' = y \in X$. Thus, the assumption that $n \not\subset X$ yields the contradiction that both $y \in X$ and $y \notin X$. Therefore $n \subset X$.

q.e.d.

Theorem 5. If X is a hereditary set such that $\phi \in X$, then $N \subset X$.

proof: Let $n \in N$; we prove $n \in X$. If $n = 0 = \phi$, then by hypothesis $n \in X$. If $n \neq \phi$, n is a successor, say $n = m' = m \cup \{m\}$, whence $m \in n$. But by the preceding lemma, $n \subset X$, and therefore $m \in X$. Since X is hereditary, $m \in X$ implies that $m' = n \in X$. Consequently, $N \subset X$.

q.e.d.*

The next theorem is very important and is known by a special name, the "Principle of Finite Induction." This is the principle referred to in high-school and freshman college texts as "Mathematical Induction." However, the name given here is appropriate, since there is another induction principle in mathematics called the "Principle of Transfinite Induction." The latter principle will not be used in this text, but the former will occur time and again as a tool for the construction of proofs.

*Theorem 6. *(Principle of Finite Induction).* If M is a hereditary subset of N, and $0 \in M$, then $M = N$.

proof: The hypotheses of the theorem are:

> (i) M is a hereditary set containing the element 0; and
> (ii) M is a subset of the set N of natural numbers.

The conclusion is that $M = N$. In order to establish this conclusion, we need only show that $M \subset N$ and $N \subset M$.

But, the hypothesis (ii) asserts $M \subset N$. And on the other hand, by hypothesis (i), $0 \in M$ and M is hereditary.   Therefore, by Theorem 5, $N \subset M$.   Hence, $M = N$.

q.e.d.*

The Principle of Finite Induction occurs in several forms, of which the statement of Theorem 6 is one.   In the course of this text, we shall state and deduce the other forms as consequences of Theorem 6.   At a small risk of confusion, all the forms will be called *the* Principle of Finite Induction.   For convenience, we shall abbreviate the expression "Principle of Finite Induction" by "PFI."

The PFI may not look very much like the statements of mathematical induction found in elementary algebra books.   Nevertheless, it is a precise formulation of what is intended by the statements of mathematical induction that the reader may be accustomed to.   The simplicity of the PFI occasionally leads one to attribute more to it than it actually says; this is a danger that the reader should guard against.

How is the PFI used in proofs?   The scheme is the following:

One wishes to prove that all the natural numbers possess a certain property.   One defines

$$M = \{x \mid x \text{ is a natural number and } x \text{ has the given property}\}.$$

Thus, M is the set consisting of all elements x such that

(a) x is a natural number, and
(b) x has the given property.

By requirement (a), it follows that $M \subset N$.   Now, if we can prove that $M = N$, then for each $y \in N$, it will be true that $y \in M$.   Hence, by (b), it must follow that every natural number has the given property. Since we already know that $M \subset N$, it suffices to prove that $0 \in M$ and M is hereditary.   For, if this is the case, then by Theorem 6, the Principle of Finite Induction, it will follow that $M = N$.

We illustrate the foregoing discussion with

* Theorem 7.   Each element of a natural number is a natural number. In other words, if $n \in N$ and if $x \in n$, then $x \in N$.

proof: We wish to show that for each natural number n it is true that

$$\text{if } x \in n \text{ then } x \in N.$$

In accordance with our scheme, we set

$$M = \{n \mid n \in N \text{ and if } x \in n \text{ then } x \in N\}.$$

By the definition of M, it follows that $M \subset N$. In order to prove that $M = N$, it suffices to show that $0 \in M$ and M is hereditary; for, if this is the case, then the PFI assures us that $M = N$.

We verify, first, that $0 = \phi \in M$. On the one hand, 0 is a natural number (by Theorem 4) and on the other, every element in 0 is a natural number (every element in 0 is a purple cow, for that matter). Thus, 0 satisfies the two requirements to be an element in M; therefore $0 \in M$.

Next, we show that if $m \in M$, then also $m' \in M$, hence M is hereditary. Now, since m is a natural number, m' is also; therefore m' satisfies the first requirement for membership in M. To prove that m' satisfies the second requirement, we must show:

$$\text{if } y \in m' \text{ then } y \in N.$$

If $y \in m' = m \cup \{m\}$ then either $y \in m$ or $y \in \{m\}$. In the latter case, $y = m$, and so y is a natural number. In the former case, $y \in m$ and $m \in M$; therefore, by definition of M, y is a natural number, i.e., $y \in N$.

Thus, M is hereditary, and by the PFI, $M = N$.

q.e.d.*

EXERCISES

Wherever appropriate, use the PFI in the following:

1. Prove that 0 is an element of every nonempty natural number. (Hint: first state clearly the desired property and then define the set M.)

2. Show that if m is a natural number, $m \neq 0$ and $m \neq 1$, then $1 \in m$.

3. Prove: If m and n are natural numbers and if $m' = n'$, then $m = n$.

4. Prove: If m is a natural number, then $m' \neq m$.

5. Let n be a natural number. Prove that if x, y, z are elements in n such that $x \in y$ and $y \in z$, then $x \in z$ (more briefly, $\in$-order is *transitive* in every natural number).

6. Prove: If x, n are natural numbers and if $x \in n$, then $x' \in n'$.

2.2.   THE ORDERING OF THE NATURAL NUMBERS

The theory of natural numbers as developed thus far has provided us with the existence of the natural numbers as sets which satisfy a rather complicated list of conditions.   It would be inconvenient to have to deal directly with these conditions throughout the rest of our work. This section is devoted to determining enough properties of the natural numbers to serve us in the future.   The most important of these properties concerns the order of the natural numbers; that is, they concern the notions of larger and smaller.   Thus we would like to say of any two distinct natural numbers that one is larger than the other.   We would like to assert that there is no natural number distinct from n and n' between n and n'.   Once several results of this type have been proved, we shall be in a position never to have to refer back to the original definitions.   Before we introduce the concept of order, we prove several results of a technical character to clear the decks.

* **Definition** ͨ.   Let X be a set.   An element $z \in X$, such that for each $y \in X$, either $y \in z$ or $y = z$ is a *final element of X*.

Theorem 8.   If $n \in N$ and if X is a nonempty subset of n, then X has a final element.

proof: Py induction.   Let M be the set of natural numbers m such that each nonempty subset of m has a final element.   Thus

$$M = \{m \mid m \in N \text{ and if } X \subset m \text{ where } X \neq \phi \text{ then } X \text{ has a final element}\}.$$

Clearly $M \subset N$.   To prove that $M = N$, we show that $0 \in M$ and M is hereditary.

Each nonempty subset of 0 has a final element (0 has no nonempty subsets), therefore $0 \in M$.

Suppose $m \in M$; we prove that $m' \in M$.   To this end, let X be a nonempty subset of m'; our object is to show that X has a final element. We distinguish two cases:

case 1: $m \notin X$.   Since $m' = m \cup \{m\}$ it follows that $X \subset m$.   Since $m \in M$, X must have a final element (definition of M).

case 2: $m \in X$.   In this case, it follows that m, itself, is a final element of X.   For if $y \in X$, then since $X \subset m'$, it follows that $y \in m' = m \cup \{m\}$.   Hence $y \in m$ or $y \in \{m\}$ and consequently $y \in m$ or $y = m$.

Cases 1 and 2 show that Definition 4 holds. Summing up, M is hereditary, $0 \in M$ and by the PFI, $M = N$.

q.e.d.

corollary: If $n \neq 0$ is a natural number then n has a final element.

proof: Take $X = n$ in Theorem 8.

Theorem 9. If $n \in N$ and $X \subset n$ and if x and y are leading elements of X, then $x = y$. Likewise, if x and y are final elements of X, then $x = y$. (This theorem justifies speaking of *the* leading element and *the* final element of a nonempty subset of a natural number.)

proof: n is $\in$-ordered. If x and y were distinct leading elements of X, then x and y would be elements of n such that both $x \in y$ and $y \in x$. This contradicts the notion of $\in$-order. The proof for final elements is similar.

q.e.d.

Theorem 10. Suppose $n \in N$, and $x \in n$. Then either x is the final element of n or $x' \in n$.*

The type of reasoning used in the proof of Theorem 10 is slightly tricky and will occur in other theorems. We therefore digress for a moment to consider this reasoning.

Suppose we are given two sentences and we wish to prove that (at least) one of them is true. The alternatives are:

1. both sentences are true;
2. one sentence is false and the other is true;
3. both sentences are false.

In view of our objective, the third alternative must be ruled out; the other two are obviously acceptable. Let us denote the sentences by "p," "q," respectively, and argue as follows:

The sentence p is either true or false and exactly one of these must hold.

(A) Suppose p is true; then surely 3 is excluded and we need go no further.

(B) Suppose p is false and *suppose that from the falsity of p we can* *deduce the truth of q.* In this case 3 is again excluded. For, otherwise, we would have

$$p \text{ is false and } q \text{ is true,}$$
$$p \text{ is false and } q \text{ is false.}$$

Hence, q would be both false and true.[6]

Thus, in a proof using this mode of reasoning, the crucial step is to deduce from the falsity of one of the statements the truth of the other. Let us see how this works out in the

*proof of theorem 10: The hypotheses of the theorem are

$$n \in N \text{ and } x \in n,$$

and we wish to prove that at least one of the two statements

$$x \text{ is the final element of } n,$$

$$x' \in n,$$

is true.

If x is the final element of n, then we are finished (see (A), page 90).

Suppose x is not the final element of n; according to our discussion (B) we prove that $x' \in n$. Since $x \in n$, $n \neq 0$, and by the corollary to Theorem 8, n has a final element y. Since x is not final, $x \in y$. Hence (Exercise 6, page 88) $x' \in y'$. But $y \in n$ yields $y' \subset n$ (Exercise 2, page 82); hence, $x' \in y' \subset n$ and $x' \in n$.

This completes the proof of Theorem 10.

Theorem 11. If x is the final element of a natural number n then $x' = n$.

proof: Since $x \in n$, as in Theorem 10 we deduce $x' \subset n$. On the other hand, for each $y \in n$, since x is final, either $y \in x$ or $y = x$. This means $y \in \{x\}$. So, for each $y \in n$, we have $y \in x \cup \{x\} = x'$. Hence, $n \subset x'$ (definition of subset). This result, together with $x' \subset n$, yields $x' = n$.

q.e.d.

Theorem 12. If $n \in N$, and if x and y are sets such that $x' = y' = n$, then x and y are natural numbers and $x = y$.

---

[6] In the Propositional Calculus, the foregoing argument is summed up by the statement " 'p or q' is equivalent with '$not\text{-}p$ implies q'."

proof: Since $x' = x \cup \{x\} = n$, it follows that $x \in n$; similarly, $y \in n$. By Theorem 7, $x \in N$ and $y \in N$. Now if $z \in n$, either $z \in x$ or $z = x$. Hence, x is a final element of n. Similarly, y is a final element of n, and by Theorem 9, $x = y$.

q.e.d.

Theorem 13. If $n \neq 0$ and $n \in N$, then there is a unique natural number m such that $m' = n$.

proof: The existence of a set m, such that $m' = n$, follows from condition (D) of Definition 7. The uniqueness follows from Theorem 12.

q.e.d.

Theorem 14. The set N of natural numbers is $\in$-ordered.

proof: Our object is to show that if m, n are natural numbers then one and only one of

(2.1) $$m \in n, \quad m = n, \quad n \in m$$

holds.

First, at most one of (2.1) holds. Indeed, if $n \in m$, then by Definition 7(C), $n \subset m$. Hence, if also $m \in n$, it would follow that $m \in m$, contrary to Theorem 2. Similarly, $n \in m$ and $m = n$ yield a contradiction.

To show that at least one of (2.1) holds, let $m \neq n$; we prove that $m \in n$ or $n \in m$ and this yields the desired result.

If $m = 0$, then $n \neq 0$ and by Exercise 1, page 88, $0 = m \in n$. Suppose $m \neq 0$ and $n \neq 0$. Then $0 = \phi \in m \cap n$, hence $m \cap n$ is a nonempty subset of a natural number (e.g., $m \cap n \subset m$). By Theorem 8, $m \cap n$ contains a final element, say x.

x is a final element of m or of n. Otherwise, since $x \in m$ and $x \in n$, it follows (Theorem 10) that $x' \in m$ and $x' \in n$, hence $x' \in m \cap n$. Since $x \in x'$ this contradicts the fact that x is the final element of $m \cap n$.

Because of the symmetry of the roles of m and n, we may assume that x is the final element of m. Then $x' = m$. But since $x \in n$, we deduce, by Theorem 10, that $x' = n$ or $x' \in n$. The first of these conclusions contradicts the hypothesis that $m \neq n$. Therefore $m = x' \in n$.

In a similar manner, if x is the final element of n, then $n \in m$.

q.e.d.

corollary 1: If m, $n \in N$, then $m \subset n$ or $n \subset m$.

proof: If $m = n$, the corollary is clearly true. If $n \in m$, then (Definition 7(C)) $n \subset m$. Similarly, if $m \in n$, then $m \subset n$.

q.e.d.

corollary 2: If m, n are natural numbers, then $m \subset n$ if and only if $m \in n$ or $m = n$. (See the Remark below.)

proof: Exercise.∗

Remark The statement of Corollary 2 contains the expression "$m \subset n$ if and only if $m \in n$ or $m = n$." Expressions of the type "p if and only if q" occur commonly in mathematics and a brief explanation of the meaning of such expressions is in order. "p if and only if q" is a short-hand way of saying "p implies q and q implies p," hence combines a statement and its converse. In greater detail:

"p only if q" means "p implies q"

and

"p if q" means "q implies p";

finally

"p if and only if q" means "p implies q and conversely, q implies p."

Sentences of the foregoing type occur not only in statements of theorems but also in definitions. To illustrate, consider Definition 5 (page 80):

1. A set A is $\in$-*ordered* means that for all x and for all y in A one and only one of the following holds:

$$x \in y, \, x = y, \text{ or } y \in x.$$

In most mathematics texts the words "means that" are replaced by "if and only if" and the resulting sentence is

2. A set A is $\in$-*ordered* if and only if for all x and all y in A one and only one of the following holds:

$$x \in y, \, x = y, \text{ or } y \in x.$$

1 and 2 are regarded as having exactly the same meaning.

As this text progresses, we shall tend, more and more, to use "if and only if" for the purpose of stating definitions.

A word of warning! It is by no means uncommon to find definitions such as 2 stated in mathematical texts as follows:

3. A set A is ∈-*ordered* if for all x and all y in A, etc.

Here, the single word "if" is being used as an abbreviation for "if and only if" and should be so understood. This usage reflects a kind of laziness afflicting many mathematicians.

EXERCISE

State Definitions 3, 4, 6, 7 using the expression "if and only if."

Once we have proved that N is ∈-ordered, we can define the usual order relation "less than" (and also "greater than") for the natural numbers. Intuitively, it is plausible that if x, y are natural numbers such that $x \in y$, then x ought to be less than y. This idea is built into

* **Definition 10.** $<$ is the subset of $N \times N$ consisting of all the ordered pairs (x,y) such that $x \in y$. More briefly:

$$< = \{(x,y) \mid (x,y) \in N \times N \text{ and } x \in y\}.$$

As is customary with binary relations, in place of "$(x,y) \in <$" we write

$$x < y,$$

and say that "x is less than y." $<$ is a *strict inequality*.

Definition 11. $\leq = \{(x,y) \mid (x,y) \in N \times N \text{ and } x < y \text{ or } x = y\}$. If $(x,y) \in \leq$ we write

$$x \leq y,$$

and say that "x is less than or equal to y." $\leq$ is a *weak inequality*.

Definition 12. $> = \{(x,y) \mid (y,x) \in < \text{ (or, } y < x)\}$;

$$\geq = \{(x,y) \mid (y,x) \in \leq \text{ (or, } y \leq x)\}.$$

In place of "$(x,y) \in >$" and "$(x,y) \in \geq$" we write

$$x > y \text{ and } x \geq y,$$

respectively. The symbols "$x > y$" and "$x \geq y$" are read "x is greater than y," and "x is greater than or equal to y," respectively.*

EXERCISES

In the following, let n, m be natural numbers.

Prove: 1. $m \leq n$ if and only if $m \subset n$.
         2. $n \leq n$ and $n \geq n$.
         3. $m \leq n$ if and only if $n \geq m$.
         4. If $n \leq m$ and $m \leq n$ then $m = n$.
         5. Either $m \leq n$ or $n \leq m$.

The simplest properties of the order relations defined above are extremely easy to verify.

* Theorem 15.   If m, n and r are natural numbers, and if $m \leq n$ and $n \leq r$ then $m \leq r$.

proof: $m \subset n$ and $n \subset r$, whence $m \subset r$.

q.e.d.

Theorem 16.   Let m, n be natural numbers.

    (a) $m < n$ if and only if $n > m$.
    (b) If $m \leq n$ then $m < n$ or $m = n$ but not both.
    (c) If $m < n$ then $m \leq n$.
    (d) One and only one of the following holds:

$$m < n, m = n, n < m.$$

proof: Exercise.

Theorem 17.   If $m < n$ and $n \leq r$ then $m < r$.
              If $m \leq n$ and $n < r$ then $m < r$.
              If $m < n$ and $n < r$ then $m < r$.

proof: We prove only the first statement and leave the rest as exercises. By Theorem 15, we know $m \leq r$. We show $m \neq r$. But, if $m = r$ then $r \leq m$. Since $n \leq r$ it follows that $n \leq m$ which, by Theorem 16, contradicts $m < n$.

Theorem 18.   If $n \in N$ then $0 \leq n$.   If $n \neq 0$ then $0 < n$.

proof: Exercise.

Theorem 19. If $n \in N$ then $n < n'$.

proof: Since $n \in n' = n \cup \{n\}$, $n < n'$ follows at once by Definition 10.

q.e.d.

Theorem 20. If $n, x \in N$ and if $n \leq x \leq n'$ then either $x = n$ or $x = n'$.

proof: The hypotheses give $n \subset x \subset n' = n \cup \{n\}$. If $n \notin x$ then $x \subset n$, so $x = n$. But if $n \in x$ then $n' = n \cup \{n\} \subset x$, so $x = n'$.

q.e.d.

corollary: There is no natural number n such that $0 < n < 1$.

Theorem 21. If $m < n$ then $m' \leq n$.

proof: If not, then $n < m'$ and then n would be a natural number between m and m' but distinct from both, in contradiction to Theorem 20.

q.e.d.*

The next theorem, the *Well-Ordering Principle* (abbreviated, WOP), is an important and powerful device for proving theorems about the natural numbers.

* **Theorem 22.** Every nonempty subset of the natural numbers contains one and only one smallest element. In other words, if $X \subset N$ and $X \neq \phi$ then there is a unique $n \in X$ such that for every $m \in X, n \leq m$.

proof: We begin with a simple remark, which the reader should verify: If $X \subset Y \subset N$ and if b is a smallest element of Y then $b \leq x$ for each $x \in X$.

The proof that every nonempty subset of N contains a smallest element proceeds by induction. Let

$$M = \{m \mid m \in N \text{ such that if } A \subset N \text{ and if } m \in A \text{ then } A$$
$$\text{contains a smallest element}\}.$$

First, $0 \in M$. For, if $0 \in A \subset N$ then $0 \leq x$ for all $x \in N$. Hence, by the remark, $0 \leq y$ for all $y \in A$. Since $0 \in A$ it follows that 0 is a smallest element in A.

Next, let $k \in M$. To prove that $k' \in M$ let $k' \in B \subset N$. If $k' \leq x$ for all $x \in B$ we are finished. If not, set

$$C = B \cup \{k\}.$$

Since $k \in C \subset N$ it follows (by the assumption that $k \in M$) that C contains a smallest element, s. Hence (by the remark), $s \leq x$ for all $x \in B$. Now, if $s \in B$ then s is a smallest element in B. If $s \notin B$ then $s = k$ and therefore $k = s < x$ for all $x \in B$. But $k' \in B$ and since there is no natural number z such that $k < z < k'$ we deduce $k' \leq x$ for all $x \in B$. Therefore k' is a smallest element in B, hence $k' \in M$. And now it follows that every nonempty subset of N has a smallest element.

Finally, every subset of N has at most one smallest element. For, if s_1, s_2 are smallest elements of $A \subset N$, then $s_1 \leq s_2$ and $s_2 \leq s_1$, whence $s_1 = s_2$.

Definition 13. If $a, b \in N$, $\llbracket a,b \rrbracket = \{x \mid x \in N \text{ and } a \leq x \text{ and } x \leq b\}$. $\llbracket a,b \rrbracket$ is the *interval from a to b*.

EXAMPLE

We illustrate the use of the WOP by proving:

For all natural numbers n, $\llbracket 0,n \rrbracket = n'$.

proof: Suppose the statement is not true. Then there are natural numbers m for which the statement is false, i.e., the set

$$M = \{m \mid m \in N \text{ and } \llbracket 0,m \rrbracket \neq m'\}$$

is a nonempty subset of the natural numbers. By the WOP, M contains a smallest element s; we emphasize that $\llbracket 0,s \rrbracket \neq s'$ since $s \in M$.

Now $0 \notin M$ since $\llbracket 0,0 \rrbracket = \{0\} = 0'$; therefore, since $s \in M$, $s \neq 0$. Consequently (by (D) of Definition 7) $s = t'$ for some natural number t. Further, $t < t'$ (by Theorem 19) so that $t \notin M$. Hence, $\llbracket 0,t \rrbracket = t'$. From this equation we obtain

$$\llbracket 0,t \rrbracket \cup \{t'\} = t' \cup \{t'\};$$

but $\llbracket 0,t \rrbracket \cup \{t'\} = \llbracket 0,t' \rrbracket$ (why?) $= \llbracket 0,s \rrbracket$ whereas $t' \cup \{t'\} = (t')' = s'$ so that $\llbracket 0,s \rrbracket = s'$, a contradiction. Therefore the assertion $\llbracket 0,n \rrbracket = n'$ is true for all $n \in N$.

· q.e.d.∗ ·

We shall have other occasions to use the WOP in the development of the theory.

EXERCISES

1. Prove $0 < 1, 0 < 2, 4 < 7, 8 < 9$.

2. Prove that if $a \leq 2$ then $a = 0$ or $a = 1$ or $a = 2$.

3. Prove that if $a < 4$ then $a = 0$ or $a = 1$ or $a = 2$ or $a = 3$.

4. Prove that if $a \leq b$, $b \leq c$, $c \leq d$, $d \leq a$, then $a = b = c = d$.

5. Let $A \subset N$; prove that $z \in A$ is a final element of A if and only if $x \leq z$ for each $x \in A$.

6. Using the PFI prove that $\mathcal{C}0,n\mathcal{I} = n'$.

7. Prove that there is a one-one correspondence between n and $\mathcal{C}1,n\mathcal{I}$. (It is possible to give proofs using the PFI and the WOP.)

8. Prove that $\mathcal{C}a,b\mathcal{I} = \phi$ if and only if $b < a$. Prove that if $a \leq b$ then $a \in \mathcal{C}a,b\mathcal{I}$ and $b \in \mathcal{C}a,b\mathcal{I}$.

9. Prove that $n \in \mathcal{C}1,n\mathcal{I}$ if and only if $n \neq 0$.

10. Prove that $\mathcal{C}1,m\mathcal{I} = \mathcal{C}1,n\mathcal{I}$ if and only if $m = n$.

11. Let m, n and p be natural numbers such that $m \leq n$. Prove that $\mathcal{C}m,n\mathcal{I} \subset \mathcal{C}m,p\mathcal{I}$ if and only if $n \leq p$.

12. Let m, n and p be natural numbers such that $m \leq n$. Prove that $\mathcal{C}m,n\mathcal{I} \subset \mathcal{C}p,n\mathcal{I}$ if and only if $p \leq m$.

13. Prove: If $m < n'$ where $m, n \in N$ then $m \leq n$.

2.3. COUNTING

In daily life one is familiar with the process of counting a set of objects. For example, to count the apples in a barrel, one might provide oneself with an empty barrel, and then transfer the apples one by one from the given barrel to the originally empty barrel. As each apple is transferred, one recites the name of a natural number. The numbers are recited in order beginning with 1. More precisely, 1 is the number recited for the first apple, and then if a is the number recited with a certain apple, a' is recited for the next apple.   One continues in this fashion until the given barrel is empty, and then announces the result that there is a certain number of apples, say 8.   Now several things have happened; in the first place, the set of numbers which has been recited in the process is the set we have denoted by " $⊏1,8⊐$ ", i.e., the set of natural numbers from 1 to 8 inclusive.   But more than this has happened; each number in $⊏1,8⊐$ has been recited in conjunction with just one apple and each apple has had just one number recited for it.   In more mathematical terms: In the course of counting, one has established a one-one correspondence between the set $⊏1,8⊐$ and the set of apples.   It is true that this one-one correspondence is lost as soon as the apples are mixed together again, but it could be preserved, if desired, by attaching labels to the apples.   These considerations suggest that the correct formal analogue of counting a set A is a one-one correspondence between A and a subset $⊏1,n⊐$ of N.

Definition 14.   A *count* of a set A is a one-one correspondence $\varphi\colon ⊏1,n⊐ \longrightarrow A$ between $⊏1,n⊐$ and A, where $n \in N$.*

There may exist sets for which no count is possible.   These sets are called infinite sets, but this matter will not concern us for the present. Unless a set is empty or has only a single element, there are several different counts of that set.   One usually expresses this fact by saying that the set can be counted in various orders.   Now if one counts a set A, one defines a one-one correspondence φ between $⊏1,n⊐$ and A and says that the result of the count is n.

Definition 15.   If $\varphi\colon ⊏1,n⊐ \longrightarrow A$ is a count of A, n is the *result* of the count φ.*[7]

Now here a very remarkable situation arises.   Referring again to the barrel of apples, the result of the count was 8.   But suppose we count the apples again.   We will presumably not count them in the same order.   Nevertheless, the result of the count will again be 8.   This fact is so familiar that it is easy to overlook its remarkable character.   *The result of a count of a set A depends only on the set A, and not on the particular count used to obtain the result.*   It is for this reason alone that it makes any sense to say that the barrel has 8 apples in it.

---

[7] We are justified in saying "the" result rather than "a" result because of the theorem that " $⊏1,n⊐ = ⊏1,m⊐$ if and only if $m = n$," Exercise 10, page 98.

In this section, our object is to give a rigorous proof of the fact that the result of two counts is always the same. The crucial part of the proof is the following theorem:

* Theorem 23. If $m, n \in N$ and if $\varphi: \mathbb{C}1,m\mathbb{J} \longrightarrow \mathbb{C}1,n\mathbb{J}$ is a one-one correspondence between $\mathbb{C}1,m\mathbb{J}$ and $\mathbb{C}1,n\mathbb{J}$, then $m = n$.

proof: The proof could be based directly on the Principle of Finite Induction, but instead we shall employ Theorem 22, the WOP, to illustrate how it can be used as a substitute. We define M as the set of natural numbers m such that there is a natural number n with $m \neq n$, and such that there is a one-one correspondence $\varphi: \mathbb{C}1,m\mathbb{J} \longrightarrow \mathbb{C}1,n\mathbb{J}$. Thus,

$$M = \{m \mid m \in N \text{ and there exists } n \in N, n \neq m, \text{ such that } \\ \text{there is a one-one correspondence } \varphi: \mathbb{C}1,m\mathbb{J} \longrightarrow \mathbb{C}1,n\mathbb{J}\}.$$

Clearly, we must prove that $M = \phi$. For, if it is indeed true that $M = \phi$, then for no natural number m can there be a one-one correspondence $\varphi: \mathbb{C}1,m\mathbb{J} \longrightarrow \mathbb{C}1,n\mathbb{J}$ with $m \neq n$. Hence, for all $m \in N$, it must be true that if there exists a one-one correspondence between $\mathbb{C}1,m\mathbb{J}$ and $\mathbb{C}1,n\mathbb{J}$ then $m = n$. We shall suppose, then, that $M \neq \phi$ and be led to a contradiction.

Since we assume $M \neq \phi$, by the WOP there is a smallest element s in M. By the definition of M, there is a natural number n and a one-one correspondence $\varphi: \mathbb{C}1,s\mathbb{J} \longrightarrow \mathbb{C}1,n\mathbb{J}$ where $n \neq s$. Now $s \neq 0$; for, if $s = 0$ then $\mathbb{C}1,s\mathbb{J} = \mathbb{C}1,0\mathbb{J} = \phi$ by Exercise 8, page 98. Then, since φ is a one-one correspondence, $\mathbb{C}1,n\mathbb{J} = \phi$; hence, by the same exercise, $n = 0$. So if $s = 0$, then $s = n$, which is not the case. By exactly the same reasoning, $n \neq 0$, thus $s \neq 0$ and $n \neq 0$. Therefore, by Exercise 9, page 98, $s \in \mathbb{C}1,s\mathbb{J}$, $n \in \mathbb{C}1,n\mathbb{J}$. Since we have shown that $s \neq 0$ and $n \neq 0$, by condition (D) in the definition of natural number there exist natural numbers a and b such that $s = a'$ and $n = b'$. By Theorem 19, $a < s$; therefore, since s is the smallest element in M, $a \notin M$. Also, $a \neq b$, since $a' \neq b'$. Now,

$$\mathbb{C}1,s\mathbb{J} = \mathbb{C}1,a'\mathbb{J} = \mathbb{C}1,a\mathbb{J} \cup \{a'\} = \mathbb{C}1,a\mathbb{J} \cup \{s\},$$

and similarly,

$$\mathbb{C}1,n\mathbb{J} = \mathbb{C}1,b\mathbb{J} \cup \{n\}.$$

By Exercise 6, page 60 (Chapter 1), there is a one-one correspondence $\eta: \mathbb{C}1,s\mathbb{J} \longrightarrow \mathbb{C}1,n\mathbb{J}$ such that $\eta(s) = n$. Define ψ by

$$\psi = \eta - \{(s,n)\}.$$

(Remember, η is a set of ordered pairs.) Then it is easy to verify (exercise!) that ψ is a one-one correspondence between $\sqsubset 1,a \sqsupset$ and $\sqsubset 1,b \sqsupset$. But, since $a \neq b$, this means that $a \in M$ (see the definition of M). Now we have a contradiction; for it was proved above that $a \notin M$. Therefore $M = \phi$.

q.e.d.

Theorem 24. If φ and ψ are two counts of a set A, whose results are m and n, respectively, then $m = n$.

proof: $\varphi \colon \sqsubset 1,m \sqsupset \longrightarrow A$ and $\psi \colon \sqsubset 1,n \sqsupset \longrightarrow A$ are one-one correspondences. Then $\psi^{-1} \circ \varphi \colon \sqsubset 1,m \sqsupset \longrightarrow \sqsubset 1,n \sqsupset$ is a one-one correspondence. Hence, by Theorem 23, $m = n$.

q.e.d.

Definition 16. If n is the result of a count of A, then A *has n elements*, or *the number of elements in A is n*, or the *cardinality of A is n*. We also denote the cardinality of A by "$\#(A)$."*

2.4.   FINITE SETS

Our theory has now progressed sufficiently that we can give a precise definition of a finite set; namely, a finite set is one for which a count exists. Our object is to prove rigorously the intuitively clear results that the union of two finite sets is finite, that a subset of a finite set is finite, and similar theorems. We will also define infinite sets simply as those which are not finite.

* **Definition 17.** A set A is *finite* if and only if there exists a count of A.

Definition 18. A set A is *infinite* if and only if it is not finite.

Theorem 25. ϕ is finite and $\#(0) = 0$.

proof: $\varphi \colon \sqsubset 1,0 \sqsupset \longrightarrow \phi$ is a count of $\phi = 0$.

q.e.d.*

EXERCISES

Prove:

1. If $\#(A) = n$ and if there is a one-one correspondence between A and B, then $\#(B) = n$.

2. If there is a one-one correspondence between A and B and if A is finite, then B is finite.

3. N is infinite.

4. If $\#(A) = n$ and if $x \notin A$, then $\#(A \cup \{x\}) = n'$. Hence, if A is finite, then $A \cup \{x\}$ is finite.

5. If $\#(A) = n$ and if $x \in A$, then $\#(A - \{x\}) = m$ where $m' = n$. Hence, if A is finite so is $A - \{a\}$.

6. A subset A of N, $A \neq \phi$, is finite if and only if it has a final element.

* Theorem 26. If A and B are finite, so is $A \cup B$.

proof: Let M be the set of natural numbers m such that if $\#(A) = m$ and if B is finite, then $A \cup B$ is finite; i.e.,

$$M = \{m \mid m \in N \text{ and if } \#(A) = m, \text{ then}$$
$$A \cup B \text{ is finite for all finite sets } B\}.$$

We prove that $M = N$. In the first place, $0 \in M$; for, if A has 0 elements, $A = \phi$, so that for all B, $A \cup B = B$. Hence, if B is finite, so is $A \cup B = B$.

Next, suppose $n \in M$ and let A be a set with n' elements. If $x \in A$, then (Exercise 5, above) $\#(A - \{x\}) = n$ and therefore $(A - \{x\}) \cup B$ is finite. But $A \cup B = ((A - \{x\}) \cup B) \cup \{x\}$, and by Exercise 4 (above) $A \cup B$ is finite.

q.e.d.

Theorem 27. If A is finite and if $B \subset A$ then B is finite.

proof: Let M be the set of natural numbers m such that for all sets A, with m elements, every subset of A is finite; thus,

$$M = \{m \mid m \in N \text{ and if } \#(A) = m \text{ then every subset of } A \text{ is finite}\}.$$

First, $0 \in M$; for, if A has 0 elements, then $A = \phi$ and the only subset of A is the finite set ϕ. Now suppose $n \in M$. To show $n' \in M$, let A be a set with n' elements and let $B \subset A$. We prove that B is finite. If $B = A$, we are finished. Suppose $B \neq A$. Then there is an element $x \in A$ such that $x \notin B$. Hence, $B \subset A - \{x\}$. But, $A - \{x\}$ contains n elements (Exercise 5) and therefore, by the assumption that $n \in M$, it follows that B is finite. Hence, $n' \in M$ and the theorem is proved for all finite sets A.

q.e.d.*

Theorem 28 is a generalization of Theorem 26.

* Theorem 28.   Suppose

> 1. A is a set;
> 2. $C \subset p(A)$;
> 3. C is finite;
> 4. for each set $B \in C$, B is finite.

Then $\cup\, C$ is finite.

In other words, a union of a finite collection of sets, each of which is finite, is finite. (More briefly: A finite union of finite sets is finite.)*

The meaning of the first formulation of Theorem 28 is illustrated by the following example:

Let $A = \{0,1,2,3,4,5\}$; then

$$p(A) = \{\phi, \{0\}, \{1\}, \ldots, \{0,1\}, \{0,2\}, \ldots, \{0,1,2,3,4,5\}\}.$$

Further, let

$$C = \{\{1\}, \{2,3\}, \{2,4\}, \{1,2,3\}\}.$$

Then $C \subset p(A)$ and $\cup\, C = \{1,2,3,4\}$.

The second formulation can be illustrated by taking the same set C as above and choosing for A any set containing the set

$$S = \{x \mid x \in B \text{ for some } B \in C\}.$$

Thus, the set A of the example above will do, but so will the sets $\{1,2,3,4\}$, $\{0,1,2,3,4\}$, $\{1,2,3,4,5\}$, as well as many others. For any choice of A (containing the set S) it is clear that $C \subset p(A)$ and that $\cup\, C = \{1,2,3,4\}$ as before.

*proof of theorem 28:   By induction on $\#(C)$. Let A be any set and let

$M = \{m \mid m \in N$ such that if $C \subset p(A)$ and $\#(C) = m$ and each element in C is a finite set, then $\bigcup C$ is finite$\}$.

First, $0 \in M$. Indeed, if $\#(C) = 0$, then $C = \phi$ and $\bigcup C = \bigcup \phi = \phi$, a finite set.

Suppose $n \in M$. To prove that $n' \in M$ let C' be a subset of $p(A)$ such that $\#(C') = n'$ and each element in C' is a finite set. Let $X \in C'$ and set $C = C' - \{X\}$. Then $\#(C) = n$ where $C \subset p(A)$ and each element in C is a finite set. Since we have assumed $n \in M$, it follows that $\bigcup C$ is finite. But $\bigcup C' = (\bigcup C) \cup X$ where $\bigcup C$ and X are both finite. By Theorem 26, $\bigcup C'$ is finite, and so $n' \in M$. Hence $M = N$ and the theorem is proved.

q.e.d.∗

EXERCISES

1. Prove Theorem 26 using the WOP. (Hint: Let $M = \{m \mid m \in N$ and if $\#(A) = m$ then there exists a finite set B such that $A \cup B$ is infinite$\}$.)

2. Prove Theorem 27 using the WOP.

* Theorem 29. Suppose that A is a finite set with m elements, and suppose that $B \subset A$. Let n be the number of elements in B. Then $n \leq m$.

proof: Let M be the set of natural numbers m such that for each set A, with m elements, and for each subset $B \subset A$, the number of elements in B is less than or equal to m, i.e.,

$M = \{m \mid m \in N$ and if $\#(A) = m$ and $B \subset A$, then $\#(B) \leq \#(A)\}$.

We prove by induction that $M = N$. First, $0 \in M$; for if A has 0 elements, $A = \phi$, and if $B \subset A$, then $B = \phi$, so B has 0 elements and $0 \leq 0$. Now, suppose $k \in M$, and let A be a set with k' elements, and let B be a subset of A.

If $B = A$, then $\#(B) = k' \leq k' = \#(A)$ and the theorem is proved. If $B \neq A$, then there is an $x \in A$ such that $x \notin B$, hence $B \subset A - \{x\}$. But, $\#(A - \{x\}) = k$ (Exercise 5, page 102) and since we have assumed $k \in M$, it follows that $\#(B) \leq \#(A - \{x\}) = k < k'$. Therefore $k' \in M$ and by the PFI, $M = N$. The proof of Theorem 29 is complete.

q.e.d.∗

EXERCISES

1. Suppose that f is a function and that $\mathfrak{D}(f)$ is finite; prove that $\mathfrak{R}(f)$ is finite. (Hint: argue by induction on the number of elements in $\mathfrak{D}(f)$. The crucial step consists in showing how, if f is a function such that $\mathfrak{D}(f)$ has m' elements, to construct a new function $\bar{f} \subset f$ such that $\mathfrak{D}(\bar{f})$ has m elements.)

2. Improve Theorem 29 to the following sharper result: If A has m elements, B has n elements, and if $B \subset A$ then $B \neq A$ if and only if $n < m$.

3. If f is a function such that $\mathfrak{D}(f)$ contains m elements, and $\mathfrak{R}(f)$ contains n elements, then $n \leq m$. Further, show that the mapping $f:\mathfrak{D}(f) \longrightarrow \mathfrak{R}(f)$ is a one-one correspondence if and only if $m = n$.

* Theorem 30. (*The pigeon-hole principle.*)   Suppose $f: A \longrightarrow B$ and that A and B are finite sets with m and n elements, respectively. Suppose also that $n < m$. Then there are elements x and $y \in A$ such that $x \neq y$, but $f(x) = f(y)$.

proof: Since $f: A \longrightarrow B$ is a mapping, we have $\mathfrak{D}(f) = A$ and $\mathfrak{R}(f) \subset B$. Let r be the number of elements in the set $\mathfrak{R}(f)$. Then, by Theorem 29, $r \leq n$, and since $n < m$, we have $r < m$. Therefore, $r \neq m$ and the mapping $f:\mathfrak{D}(f) \longrightarrow \mathfrak{R}(f)$ is not a one-one correspondence (Exercise 3, above). The conclusion of the theorem is now immediate from the definition of a one-one correspondence.

q.e.d._*

Theorem 30 is called "the pigeon-hole principle," since it is the mathematical formulation of the idea that if one places m objects in n pigeon holes, where $n < m$, then one must place at least two objects in one of the pigeon holes. Think of A as a set of objects, B as the set of pigeon holes, and for $x \in A$ think of $f(x)$ as the hole into which the object x is placed.

* Theorem 31.   If A and B are finite sets, then $A \times B$ is finite.

proof: For each element $x \in A$, define B_x as the set of all ordered pairs (x,y) such that $y \in B$. Define the mapping $\varphi_x: B \longrightarrow B_x$ by $\varphi_x(y) = (x,y)$. It is easy to check that φ_x is a one-one correspondence. Suppose $f: \mathbf{[}1,m\mathbf{]} \longrightarrow B$ is a count of B. Then $\varphi_x \circ f: \mathbf{[}1,m\mathbf{]} \longrightarrow B_x$ is a one-one correspondence, since it is a composite of two one-one correspondences. Hence $\varphi_x \circ f \colon \complement 1,m \complement \longrightarrow B_x$ is a count of B_x, and B_x is a finite set. Now let $\mathfrak{B}$ be the set whose elements are the sets B_x where $x \in A$. The mapping $\psi \colon A \longrightarrow \mathfrak{B}$ defined by $\psi(x) = B_x, x \in A$, is a one-one correspondence ($x \neq y$ if and only if $B_x \neq B_y$). Then if we compose ψ with a count of A, we obtain a count of $\mathfrak{B}$; therefore $\mathfrak{B}$ is finite. Now $\mathfrak{B} \subset p(A \times B)$, and is a finite set whose elements are finite sets. Therefore $\bigcup \mathfrak{B}$ is a finite set (Theorem 28). But $\bigcup \mathfrak{B} = A \times B$. For, since $\mathfrak{B} \subset p(A \times B)$, $\bigcup \mathfrak{B} \subset A \times B$. Also, if $(x,y) \in A \times B$, then $(x,y) \in B_x$, and since $B_x \in \mathfrak{B}$, $(x,y) \in \bigcup \mathfrak{B}$. Hence $A \times B \subset \bigcup \mathfrak{B}$. Therefore $A \times B$ is finite.

q.e.d._*

2.5. ADDITION AND MULTIPLICATION

Now we are in a position to define addition and multiplication for the elements of N. These definitions are based on the usual intuitive notions of these operations for natural numbers. Consider addition: if m and n are two natural numbers, how does one find $m + n$? One takes two disjoint sets, A and B, with m and n elements respectively, and counts $A \cup B$. Then the number of elements in $A \cup B$ is $m + n$. This is, in fact, what the grade school child does when he counts upon his fingers to assist him in carrying out simple additions. We use a device based on this idea, in order to define $m + n$. Given m and n, we might be tempted to take $\complement 1,m \complement$ and $\complement 1,n \complement$ as the sets A and B respectively. However, these sets are not in general disjoint, and therefore will not do for our purposes. On the other hand, the sets $\complement 1,m \complement \times \{0\}$ and $\complement 1,n \complement \times \{1\}$ contain m and n elements respectively, and are disjoint. For $\complement 1,m \complement \times \{0\}$ is the set of ordered pairs $(x,0)$, $x \in \complement 1,m \complement$ and $\complement 1,n \complement \times \{1\}$ is the set of ordered pairs $(x,1)$ with $x \in \complement 1,n \complement$. Since $1 \neq 0$, these two sets are disjoint. Then, because of the results of Sections 2.3 and 2.4, $(\complement 1,m \complement \times \{0\}) \cup (\complement 1,n \complement \times \{1\})$ is a finite set with a well-determined number of elements, and this number is by definition $m + n$. We proceed to the formal development.

* Definition 19. If $m, r \in N$, $m_r = \complement 1,m \complement \times \{r\}$.

In particular, $m_0 = \complement 1,m \complement \times \{0\}$ and $m_1 = \complement 1,m \complement \times \{1\}$.

Theorem 32. If $m, r \in N$, then m_r is finite and has m elements.

proof: Exercise.

Theorem 33. If $m, n, r, s \in N$ and $r \neq s$, then $m_r \cap n_s = \phi$.

proof: Exercise.

Theorem 34. If $m, n, r, s \in N$, then $m_r \cup n_s$ is finite.

proof: Theorem 26.

Definition 20. If $m, n \in N$, then the *sum of m and n*, $m + n$, is the number of elements in $m_0 \cup n_1$.

Theorem 35. If A and B are finite sets with m and n elements, respectively, and if $A \cap B = \phi$, then $A \cup B$ has $m + n$ elements.

proof: Let $f : \mathbb{C} 1, m \mathbb{J} \longrightarrow A$ be a count of A and let $g : \mathbb{C} 1, n \mathbb{J} \longrightarrow B$ be a count of B. Let $\varphi : m_0 \longrightarrow \mathbb{C} 1, m \mathbb{J}$ be defined by $\varphi((x,0)) = x$ and let $\psi : n_1 \longrightarrow \mathbb{C} 1, n \mathbb{J}$ be defined by $\psi((x,1)) = x$. These last two mappings are one-one correspondences; hence $f \circ \varphi : m_0 \longrightarrow A$ and $g \circ \psi : n_1 \longrightarrow B$ are one-one correspondences. Since $m_0 \cap n_1 = \phi$ and $A \cap B = \phi$, it follows that $(f \circ \varphi) \cup (g \circ \psi) : m_0 \cup n_1 \longrightarrow A \cup B$ is a one-one correspondence. Let $w = (f \circ \varphi) \cup (g \circ \psi)$. By the definition of $m + n$ there is a count $h : \mathbb{C} 1, m + n \mathbb{J} \longrightarrow m_0 \cup n_1$. Then $w \circ h : \mathbb{C} 1, m + n \mathbb{J} \longrightarrow A \cup B$ is a count of $A \cup B$.

q.e.d.

Theorem 36. If $m, n \in N$ then $m + n = n + m$. (This theorem is the *commutative law for the addition of natural numbers*.)

proof: Let A and B be disjoint sets with m and n elements, respectively, say $A = m_0$, $B = n_1$. Then by Theorem 35, $A \cup B$ has $m + n$ elements. Likewise, by Theorem 35, $B \cup A$ has $n + m$ elements. Since $A \cup B = B \cup A$, it follows that $m + n = n + m$.

q.e.d.

Theorem 37. If $m, n, r \in N$ then $(m + n) + r = m + (n + r)$. (Theorem 37 is the *associative law for the addition of natural numbers*.)

proof: Let A, B, C be sets with m, n, r elements, respectively, such that $A \cap B = A \cap C = B \cap C = \phi$. Such sets exist; for instance, $\mathbb{C} 1, m \mathbb{J} \times \{0\}$, $\mathbb{C} 1, n \mathbb{J} \times \{1\}$, $\mathbb{C} 1, r \mathbb{J} \times \{2\}$. Then by Theorem 35, $A \cup B$ has $m + n$ elements. Now $(A \cup B) \cap C = \phi$, for

$(A \cup B) \cap C = (A \cap C) \cup (B \cap C) = \phi \cup \phi = \phi$. Therefore, by Theorem 35, $(A \cup B) \cup C$ has $(m + n) + r$ elements. Likewise, $B \cup C$ has $n + r$ elements, and since $A \cap (B \cup C) = (A \cap B) \cup (A \cap C) = \phi \cup \phi = \phi$, $A \cup (B \cup C)$ has $m + (n + r)$ elements. Since

$$(A \cup B) \cup C = A \cup (B \cup C)$$

it follows that $(m + n) + r = m + (n + r)$.

q.e.d.

Definition 21. If $m, n, r \in N$, $m + n + r = (m + n) + r$.

Theorem 38. $m + n + r = (m + n) + r = m + (n + r)$.

proof: Theorem 37.

Theorem 39. $m + n + r = m + r + n = n + m + r = n + r + m = r + m + n = r + n + m$.

proof: Each of the following equalities is justified by one of Theorems 36, 37, or 38: $m + n + r = m + (n + r) = m + (r + n) = m + r + n = (m + r) + n = (r + m) + n = r + m + n = r + (m + n) = r + (n + m) = r + n + m = (r + n) + m = (n + r) + m = n + r + m = n + (r + m) = n + (m + r) = n + m + r$.

q.e.d.

Theorem 40. If $n \in N$, then $n + 0 = 0 + n = n$.

proof: Exercise.

Theorem 41. If $n \in N$, then $n' = n + 1$.

proof: $\mathsf{C}1,n'\mathsf{J} = \mathsf{C}1,n\mathsf{J} \cup \{n'\}$, $\mathsf{C}1,n\mathsf{J} \cap \{n'\} = \phi$.

q.e.d.

Definition 22. $9' = 9 + 1 = 10$.

Theorem 42. If $m, n \in N$, then $m' + n = m + n'$.

proof: By Theorem 41, $m' = m + 1$. Therefore $m' + n = (m + 1) + n$. By Theorems 37, 36 and 41, $(m + 1) + n = m + (1 + n) = m + (n + 1) = m + n'$.

q.e.d.

Theorem 43. For the values of m and n shown in the table, the value of $m + n$ can be read from the table.[8]

m \ n	0	1	2	3	4	5	6	7	8	9	10
0	0	1	2	3	4	5	6	7	8	9	10
1	1	2	3	4	5	6	7	8	9	10	$10+1$
2	2	3	4	5	6	7	8	9	10	$10+1$	$10+2$
3	3	4	5	6	7	8	9	10	$10+1$	$10+2$	$10+3$
4	4	5	6	7	8	9	10	$10+1$	$10+2$	$10+3$	$10+4$
5	5	6	7	8	9	10	$10+1$	$10+2$	$10+3$	$10+4$	$10+5$
6	6	7	8	9	10	$10+1$	$10+2$	$10+3$	$10+4$	$10+5$	$10+6$
7	7	8	9	10	$10+1$	$10+2$	$10+3$	$10+4$	$10+5$	$10+6$	$10+7$
8	8	9	10	$10+1$	$10+2$	$10+3$	$10+4$	$10+5$	$10+6$	$10+7$	$10+8$
9	9	10	$10+1$	$10+2$	$10+3$	$10+4$	$10+5$	$10+6$	$10+7$	$10+8$	$10+9$

proof: We shall not carry out the entire proof, but shall illustrate the procedure in special cases. The main idea is to use Theorem 41 to show that all the sums along a diagonal from lower left to upper right are equal. Since every such diagonal in the table terminates on the upper row or the final column, where the values of the sums follow from either Theorem 40 or Theorem 36, each entry in the table can be verified. For example:

$$
\begin{aligned}
4 + 0 &= 3' + 0 &&\text{(Definition of 4)} \\
&= 3 + 0' &&\text{(Theorem 42)} \\
= 3 + 1 &= 2' + 1 &&\text{(Definitions of 1 and 3)} \\
&= 2 + 1' &&\text{(Theorem 42)} \\
= 2 + 2 &= 1' + 2 &&\text{(Definition of 2, used twice)} \\
&= 1 + 2' &&\text{(Theorem 42)} \\
= 1 + 3 &= 0' + 3 &&\text{(Definitions of 3 and 1)} \\
&= 0 + 3' &&\text{(Theorem 42)} \\
&= 0 + 4 &&\text{(Definition of 4)} \\
&= 4 &&\text{(Theorem 40)}
\end{aligned}
$$

Summing up, $4 + 0 = 3 + 1 = 2 + 2 = 1 + 3 = 0 + 4 = 4$. In similar fashion, $9 + 5 = 8 + 6 = 7 + 7 = 6 + 8 = 5 + 9 = 4 + 10 = 10 + 4$, where the last equality is an instance of Theorem 36.∗

[8] In particular, the famous theorem "$2 + 2 = 4$" is included in these results.

The table of Theorem 43 can be written in terms of more familiar numerals if we define "11," "12," . . . , "19" appropriately. Assuming that "11" is defined by $11 = 10 + 1$, it becomes necessary to prove, say, that $6 + 5 = 11$.

We turn now to the subject of multiplication.

* Definition 23. If m, $n \in N$, the number $\#(\mathsf{C}1,m\mathsf{J} \times \mathsf{C}1,n\mathsf{J})$ is the *product* $m \cdot n$ (or mn) of m and n. ($\mathsf{C}1,m\mathsf{J} \times \mathsf{C}1,n\mathsf{J}$ is finite because of Theorem 31.)

Theorem 44. If A and B are finite sets such that $\#(A) = m$ and $\#(B) = n$, then $\#(A \times B) = mn$.

proof: By assumption, there are one-one correspondences

$$\varphi_1: \mathsf{C}1,m\mathsf{J} \longrightarrow A \text{ and } \varphi_2: \mathsf{C}1,n\mathsf{J} \longrightarrow B.$$

Define $\psi: \mathsf{C}1,m\mathsf{J} \times \mathsf{C}1,n\mathsf{J} \longrightarrow A \times B$ by $\psi((a,b)) = (\varphi_1(a),\varphi_2(b))$. It is easily verified that ψ is a one-one correspondence. On the other hand, by the definition of mn, there is a one-one correspondence $\omega: \mathsf{C}1,mn\mathsf{J} \longrightarrow \mathsf{C}1,m\mathsf{J} \times \mathsf{C}1,n\mathsf{J}$. Then $\psi \circ \omega: \mathsf{C}1,mn\mathsf{J} \longrightarrow A \times B$ is a one-one correspondence.

q.e.d.

Theorem 45. $mn = nm$. (*Commutative law of multiplication of natural numbers.*)

proof: Let A and B have m and n elements, respectively. Then $\#(A \times B) = mn$ and $\#(B \times A) = nm$. On the other hand, the mapping $\varphi: A \times B \longrightarrow B \times A$ defined by $\varphi((a,b)) = (b,a)$ is clearly a one-one correspondence.

q.e.d.

Theorem 46. $(mn)r = m(nr)$. (*Associative law of multiplication of natural numbers.*)

proof: Suppose A, B, C have m, n, r elements, respectively. Then $\#((A \times B) \times C) = (mn)r$ and $\#(A \times (B \times C)) = m(nr)$. But the mapping $\eta: (A \times B) \times C \longrightarrow A \times (B \times C)$ defined by $\eta(((a,b),c)) = (a,(b,c))$ is a one-one correspondence.

q.e.d.

Theorem 47. $(m + n)r = (mr) + (nr)$. (*Distributive law.*)

proof: Let A, B, C have m, n, r elements, respectively.   Suppose also that $A \cap B = \phi$.   (Such sets exist.   Examples?)   Then $\#(A \cup B) = m + n$.   Therefore $\#((A \cup B) \times C) = (m + n)r$.   Also

$$\#(A \times C) = mr, \quad \#(B \times C) = nr.$$

The reader may verify that $(A \times C) \cap (B \times C) = \phi$.   Therefore $\#((A \times C) \cup (B \times C)) = \#(A \times C) + \#(B \times C) = (mr) + (nr)$.   But $(A \times C) \cup (B \times C) = (A \cup B) \times C$.

q.e.d.

Theorem 48. $n \cdot 0 = 0 \cdot n = 0$.

proof: For any set A, $A \times \phi = \phi$ (Exercise 2, page 46).

q.e.d.

Theorem 49. $n \cdot 1 = 1 \cdot n = n$.

proof: $\{0\}$ has 1 element.   If A has n elements, there is a one-one correspondence between $A \times \{0\}$ and A.

q.e.d._*

Note: An expression such as "$(mr) + (nr)$" is usually written as "$mr + nr$."   We are going to assume that the reader is familiar with the conventions for omitting certain parentheses in situations such as this.

*Theorem 50.   (The multiplication table.)   For the values of m and n shown, the value of mn is given by the table:

m \ n	0	1	2	3	4	5	6	7	8	9
0	0	0	0	0	0	0	0	0	0	0
1	0	1	2	3	4	5	6	7	8	9
2	0	2	4	6	8	10	$10+2$	$10+4$	$10+6$	$10+8$
3	0	3	6	9	$10+2$	$10+5$	$10+8$	$2 \cdot 10+1$	$2 \cdot 10+4$	$2 \cdot 10+7$
4	0	4	8	$10+2$	$10+6$	$2 \cdot 10$	$2 \cdot 10+4$	$2 \cdot 10+8$	$3 \cdot 10+2$	$3 \cdot 10+6$
5	0	5	10	$10+5$	$2 \cdot 10$	$2 \cdot 10+5$	$3 \cdot 10$	$3 \cdot 10+5$	$4 \cdot 10$	$4 \cdot 10+5$
6	0	6	$10+2$	$10+8$	$2 \cdot 10+4$	$3 \cdot 10$	$3 \cdot 10+6$	$4 \cdot 10+2$	$4 \cdot 10+8$	$5 \cdot 10+4$
7	0	7	$10+4$	$2 \cdot 10+1$	$2 \cdot 10+8$	$3 \cdot 10+5$	$4 \cdot 10+2$	$4 \cdot 10+9$	$5 \cdot 10+6$	$6 \cdot 10+3$
8	0	8	$10+6$	$2 \cdot 10+4$	$3 \cdot 10+2$	$4 \cdot 10$	$4 \cdot 10+8$	$5 \cdot 10+6$	$6 \cdot 10+4$	$7 \cdot 10+2$
9	0	9	$10+8$	$2 \cdot 10+7$	$3 \cdot 10+6$	$4 \cdot 10+5$	$5 \cdot 10+4$	$6 \cdot 10+3$	$7 \cdot 10+2$	$8 \cdot 10+1$

proof: To illustrate the procedure, we calculate the multiples of 7. The various steps are justified by the theorems on multiplication or by the addition table.

$$0 \cdot 7 = 0,$$
$$1 \cdot 7 = 7,$$
$$2 \cdot 7 = (1 + 1) \cdot 7 = 1 \cdot 7 + 1 \cdot 7 = 7 + 7 = 10 + 4,$$
$$3 \cdot 7 = (2 + 1) \cdot 7 = 2 \cdot 7 + 1 \cdot 7 = (10 + 4) + 7$$
$$= 10 + (4 + 7) = 10 + (10 + 1) = (10 + 10) + 1$$
$$= (1 \cdot 10 + 1 \cdot 10) + 1 = (1 + 1) \cdot 10 + 1 = 2 \cdot 10 + 1,$$
$$4 \cdot 7 = (3 + 1) \cdot 7 = 3 \cdot 7 + 1 \cdot 7 = (2 \cdot 10 + 1) + 7$$
$$= 2 \cdot 10 + (1 + 7) = 2 \cdot 10 + 8,$$
$$5 \cdot 7 = (4 + 1) \cdot 7 = 4 \cdot 7 + 7 = (2 \cdot 10 + 8) + 7$$
$$= 2 \cdot 10 + (8 + 7) = 2 \cdot 10 + (10 + 5)$$
$$= (2 \cdot 10 + 10) + 5 = (2 + 1) \cdot 10 + 5 = 3 \cdot 10 + 5,$$

etc. *

As in Theorem 43, the table of Theorem 50 can be brought into familiar form by giving appropriate definitions of "11," "12," ..., "81." Assuming, for instance, that "55" has been defined and that one then defines "56" by $56 = 55 + 1$, it becomes necessary to prove that $5 \cdot 10 + 6 = 56$.

2.6. THE RELATIONS BETWEEN ORDER, ADDITION AND MULTIPLICATION

* Theorem 51. Suppose that m, n and r are natural numbers, and that $n < r$. Then $m + n < m + r$.

proof: There exist sets A, B and C such that

(i) $\#(A) = m$, $\#(B) = n$, $\#(C) = r$,

(ii) $A \cap C = \phi$,

(iii) $B \subset C$.

We leave the reader the exercise of exhibiting such sets. Since $B \subset C$, it follows that $A \cap B = \phi$. Therefore, from Theorem 35, we deduce that $\#(A \cup B) = m + n$ and $\#(A \cup C) = m + r$. On the other hand, $A \cup B \subset A \cup C$ and $A \cup B \neq A \cup C$. This last follows from the fact that $B \neq C$, which is in turn a consequence of $\#(B) \neq \#(C)$ and

$A \cap B = \phi$.   Exercise 2, page 105, now yields the result $\#(A \cup B) <$ $\#(A \cup C)$, i.e., $m + n < m + r$.

q.e.d.

Theorem 52. If m, n and r are natural numbers, and if $n \leq r$, then $m + n \leq m + r$.

proof: If $n \leq r$, then by Theorem 16, either $n = r$ or $n < r$. If $n = r$, then $m + n = m + r$, and hence $m + n \leq m + r$. If $n < r$, then by Theorem 51, $m + n < m + r$; hence $m + n \leq m + r$.

q.e.d.

Theorem 53. If m and n are natural numbers, then $m \leq m + n$. If $n \neq 0$, then $m < m + n$.

proof: By Theorem 18, $0 \leq n$. Hence, $m = m + 0 \leq m + n$. If $n \neq 0$, then $0 < n$ and $m = m + 0 < m + n$.

q.e.d.

Theorem 54. (*Law of cancellation for addition.*) If m, n and r are natural numbers and if $m + n = m + r$, then $n = r$.

proof: We know from Theorem 16 that either (i) $n < r$, (ii) $r < n$, or (iii) $n = r$. But if $n < r$, then by Theorem 51, $m + n < m + r$ in contradiction to the assumption that $m + n = m + r$. Likewise, if $r < n$, Theorem 51 leads to a contradiction. Possibilities (i) and (ii) being false, the remaining alternative, namely $n = r$, must be true.

q.e.d.*

Theorem 54 is the justification of the formal computational device of canceling equal terms from both sides of an equation.

* Theorem 55. If $m \leq n$, there is a natural number x such that $m + x = n$. Further, there is only one such number x.

proof: $[m',n]$ is a subset of $[1,n]$ and, hence, is finite, so that $\#([m',n])$ is defined. Let $x = \#([m',n])$. To show that $m + x = n$, we prove that (i) $[1,m] \cap [m',n] = \phi$ and (ii) $[1,m] \cup [m',n] = [1,n]$. (i) If $a \in [1,m] \cap [m',n]$, then $m' \leq a$ and $a \leq m$ so that $m' \leq m$, which is false. Hence, $[1,m] \cap [m',n] = \phi$. (ii) Since $m \leq n$, it follows that $[1,m] \subset [1,n]$.

Also, $\lceil m',n \rceil \subset \lceil 1,n \rceil$; hence, $\lceil 1,m \rceil \cup \lceil m',n \rceil \subset \lceil 1,n \rceil$. On the other hand, suppose $a \in \lceil 1,n \rceil$. Then $1 \leq a$ and $a \leq n$. Now by Theorems 16 and 21, either $a \leq m$ or $m' \leq a$. If $a \leq m$, then since $1 \leq a$, $a \in \lceil 1,m \rceil$. If $m' \leq a$, then since $a \leq n$, $a \in \lceil m',n \rceil$. In any case, $a \in \lceil 1,m \rceil \cup \lceil m',n \rceil$. Hence, $\lceil 1,n \rceil \subset \lceil 1,m \rceil \cup \lceil m',n \rceil$. The uniqueness of x follows from Theorem 54.

q.e.d.

Theorem 56. If $m \cdot n = 0$, then $m = 0$ or $n = 0$.

proof: Suppose $m \neq 0$ and $n \neq 0$. Let A, B be sets such that $\#(A) = m$, $\#(B) = n$. Since neither m nor n is zero, A and B are both nonempty. Let $a \in A$, $b \in B$; then $(a,b) \in A \times B$, so that $A \times B \neq \phi$. Therefore, $m \cdot n = \#(A \times B) \neq 0$, a contradiction. Hence $m = 0$ or $n = 0$.

q.e.d.*

In Volume II (Algebra) of this work, we shall study arithmetics in which Theorem 56 is false. There are arithmetics containing elements $x \neq 0$, $y \neq 0$ but $x \cdot y = 0$.

* Theorem 57. If m, n and r are natural numbers, if $m \neq 0$ and if $n < r$, then $mn < mr$.

proof: By Theorem 55, there is a number x such that $n + x = r$. Then $m \cdot r = m(n + x) = mn + mx$. But $m \neq 0$, and since $r > n$, $x \neq 0$. Therefore, by Theorem 56, $mx \neq 0$. By Theorem 53, $mn < mr$.

q.e.d.

Theorem 58. If m, n and r are natural numbers and if $n \leq r$, then $mn \leq mr$.

proof: Exercise.

Theorem 59. (*Law of cancellation for multiplication.*) If m, n and r are natural numbers and if $m \neq 0$ and if $mn = mr$, then $n = r$.

proof: Use Theorem 57 to show that $n < r$ and $r < n$ are both false.

q.e.d.*

It must be observed that the hypothesis "$m \neq 0$" is essential in Theorem 59. Here is a well-known "proof" that $1 = 2$. First, $0 = 0 \cdot 1$, $0 = 0 \cdot 2$; therefore $0 \cdot 1 = 0 \cdot 2$. Cancel the zeros; $1 = 2$. The fallacy lies in the misapplication of Theorem 59; namely, Theorem 59 has been applied in a case where $m = 0$. There are ways of doctoring the above argument to make it less apparent that a cancellation of zeros is taking place.

EXERCISE

Prove: if $mn < mr$ then $n < r$.

2.7.   THE PRINCIPLE OF FINITE INDUCTION, AGAIN[9]

In elementary algebra texts where induction is discussed for a first time, one finds that the inductions usually "start with one." This is in contrast with the PFI as stated in Theorem 6; all of the inductions in the present chapter have "started with zero." Actually, neither zero nor one is consecrated to the purpose of induction. Our next theorem will show that an induction may "start with" any natural number.

* Definition 24.   Let a be a natural number. Then

$$N_a = \{x \mid x \in N \text{ and } x \geq a\}.$$

Theorem 60.   If

(i) K is a nonempty, hereditary subset of N_a, and
(ii) if a is an element in K,

then $K = N_a$.

Note: If $a = 0$, then $N_0 = N$, and Theorem 60 is the original PFI.

proof: By the Well-Ordering Principle.

---

[9] To students reading a book such as this for the first time, we suggest:
(a) Skip the proofs in Section 2.7;
(b) Skip Section 2.8 until ready to begin Chapter 4;
(c) Skim Section 2.9 and review it as needed.

Suppose $K \neq N_a$. By (i), there is an element $x \in N_a$ such that $x \notin K$. Hence, the set

(2.2) $$L = \{m \mid m \in N_a \text{ and } m \notin K\}$$

is a nonempty subset of N_a (and therefore, also, of N). By the WOP L contains a smallest element s; since $s \in L$, it follows that $s \in N_a$, whence $s \geq a \geq 0$. Clearly, $s \neq a$, since $a \in K$, but $s \notin K$. Therefore $s > a \geq 0$ and it follows that $s = t'$ for some natural number t. Further, since $t < t' = s$, we deduce that $t \in K$. But $t' = s \notin K$. This is a contradiction, since it was assumed that K is hereditary. Therefore $K = N_a$.

q.e.d.$_*$

We give yet another form of the PFI, which is used in proving a variety of theorems. Indeed, we shall see that the so-called "Fundamental Theorem of Arithmetic" (see Chapter 3) requires this new form.

* Theorem 61. Let K be a nonempty subset of N_a and let a be the smallest element in K. Further, assume:

(†) If $n \in N_a$, and if $\{x \mid a \leq x < n\} \subset K$ then $n \in K$.
Then $K = N_a$.

proof: The proof uses the WOP and parallels closely the proof of Theorem 60. Define L by (2.2) and suppose, as in Theorem 60, that $K \neq N_a$. Then L is nonempty and contains a smallest element s. As before, $s \notin K$, $s \in N_a$ and $s > a$. Since s is the smallest element such that $s > a$ and $s \notin K$, it follows that for all $y \in N_a$ such that $a \leq y < s$ we must have $y \in K$. But then, by (†), $s \in K$, a contradiction. Therefore $K = N_a$.

q.e.d.$_*$

EXERCISE

Prove Theorem 60 using Theorem 6 and the mapping $\varphi : N \longrightarrow N$ defined by

$$\varphi(x) = a + x, \ x \in N,$$

where $a \in N$.

We conclude this section with a warning concerning the misuse of the PFI. By a properly careless application of the PFI or by a disregard for the meanings of words, some strange and wonderful results can be "proved." To illustrate the dangers, we give one example now and two examples at the end of Section 2.9. It is left to the reader to determine what, if anything, is wrong.

EXAMPLE

All natural numbers are equal.

"proof:" The object is to show that for all natural numbers k, if M is a subset of N containing k elements, then the elements of M are equal.

If M contains no elements, then $M = \phi$, and clearly $x = y$ for all $x, y \in M$.

Assume the proposition is true for all sets containing n elements, $n \in N$. Let S be a subset of N containing $n + 1$ elements, say, $S = \{a_1, a_2, \ldots, a_{n+1}\}$. Let $S_1 = S - \{a_1\}$, $S_2 = S - \{a_2\}$. Then S_1 is a subset of N containing n elements, hence $a_2 = a_3 = \ldots = a_{n+1}$. Similarly, S_2 is a subset of N containing n elements; therefore $a_1 = a_3 = a_4 = \ldots = a_{n+1}$. Consequently, $a_1 = a_2 = a_3 = \ldots = a_{n+1}$, as asserted.

2.8.   SEQUENCES

A particular kind of function which is essential for the further development of arithmetic is the "sequence." Recall, first, that the sets N_a, $a \in N$ were defined by

$$N_a = \{x \mid x \in N \text{ and } x \geq a\}.$$

Thus, each N_a is a subset of N, and in particular $N_0 = N$. Note that for all $a, b \in N$, there is an especially simple one-one correspondence between N_a and N_b. Without loss of generality, we may assume that $a \leq b$; then there is a natural number k such that $b = a + k$. The mapping $\zeta : N_a \longrightarrow N_b$ defined by

$$\zeta(y) = y + k, \, y \in N_a$$

is obviously one-one and onto.

* Definition 25. Let S be a set. A mapping $\varphi : N_a \longrightarrow S$ is an *infinite sequence of elements of S.*

Thus, sequences are distinguished among all functions by the fact that their domains are the sets N_a, $a \in N$.

Since, in this text, we shall have no occasion to consider sequences other than infinite, we shall henceforth omit the word "infinite" in all further discussion of sequences. Also, whenever the set S containing the range of the sequence is clear from the context, the expression "of elements of S" will be omitted. In short, whenever the set S is understood, the word "sequence" will be considered as an abbreviation for "infinite sequence of elements of S."

In most cases, the domain of the sequences which we study will be N_1, and occasionally, $N_0 = N$. Sequences having other N_a's as domains will occur very infrequently. If the domain of a sequence is not mentioned, it is to be understood that the domain is N_1; in all other cases, the domain will be given explicitly.

Definition 26. If φ is a sequence with domain N_a, then for each $n \in N_a$, $\varphi(n)$ is the *n-th term* of the sequence φ.

In other words, the n-th term of a sequence φ is the *second component* of the ordered pair $(n,\varphi(n)) \in \varphi$.

If φ is a sequence, one frequently writes "x" (or "y," or "z," etc.) in place of "φ" and denotes the n-th term by "x_n" (or "y_n," or "z_n," etc.) instead of using "$\varphi(n)$." The sequence φ itself—with domain N_1—will also be denoted by

$$[\varphi(n)], \quad \text{or: } [x_n], \quad \text{or: } [x_1,x_2,x_3, \ldots].$$

In case the domain of φ is N_a, the alternative notations for φ are

$$[\varphi(n)], \ n \in N_a, \quad \text{or: } [x_n], \ n \in N_a, \quad \text{or: } [x_a,x_{a+1},x_{a+2}, \ldots].$$

These alternative notations for φ are notations in which the second components of the ordered pairs of φ are represented.∗

A sequence of elements of S should not be confused with a subset of S despite any similarities of notation. The examples below will serve to emphasize the difference between the two concepts.

EXAMPLES

1. Let $S = N$, $\varphi(n) = 1$ for all $n \in N_1$. The notations for φ are "$[\varphi(n)]$" and "$[1]$" and "$[1,1,1, \ldots]$," and the n-th term of φ, for

each $n \in N_1$, is one.   On the other hand, the set of second components of φ is the set $\{1\}$ consisting of the single element, one. Thus, although a sequence (in this text) is always a set consisting of infinitely many elements, each element being an ordered pair, the set of second components may contain only a single element.

2. $S = Z$, the set of integers (not yet defined, see Chapter 3),

$$\varphi(n) = \begin{cases} 1, \text{ if } n \text{ is odd,} \\ -1, \text{ if } n \text{ is even.} \end{cases}$$

This sequence may be denoted by

$$[1,-1,1,-1,\ldots],$$

but the set of terms of φ, i.e., the set of second components of φ is

$$\{1,-1\}.$$

3. $S = R$, the set of rational numbers (not yet defined, see Chapter 3), $\mathfrak{D}(\varphi) = N$, and

$$\varphi(n) = \begin{cases} 1, \text{ if } n \text{ is odd,} \\ (\tfrac{1}{2})^n, \text{ if } n \text{ is even.} \end{cases}$$

The sequence may be denoted by

$$[1,1,\tfrac{1}{4},1,\tfrac{1}{16},1,\tfrac{1}{64},\ldots]$$

and the set of terms, i.e., the set of second components of the ordered pairs of φ is

$$\left\{ \frac{1}{2^{2n}} \,\middle|\, n \in N \right\}.$$

4. $S = N_1$,

$$\varphi(n) = \begin{cases} n, \text{ if } n \leq 5, \\ 6, \text{ if } n \geq 6. \end{cases}$$

A notation for this sequence is

$$[1,2,3,4,5,6,6,6,\ldots]$$

and the range of φ is

$$\{1,2,3,4,5,6\}.$$

The few facts concerning sequences, presented above, will be used at once in the study of *recursive definitions*.

2.9. RECURSIVE DEFINITIONS

In elementary algebra, the powers of a number y are defined by the equations

$$y^0 = 1,$$
$$y^1 = y,$$
$$y^2 = y \cdot y,$$
$$y^3 = (y \cdot y) \cdot y,$$
$$y^4 = ((y \cdot y) \cdot y) \cdot y,$$

"and so on." For the present, the only numbers at our disposal are the natural numbers, and so the range of y will be N. Still, with this restriction in mind, we ask, "What is meant by 'and so on'?"

Perhaps we can obtain a clearer understanding of the question by citing a definition of powers as it appears in more advanced books:

For all $y \in N$,

(2.3) $y^0 = 1$, and for all $m \in N$, $y^{m+1} = y^m \cdot y$.

As a consequence of (2.3), it is asserted: y^k is a natural number for all $y \in N$ and for all $k \in N$. We remark, by the way, that (2.3) is an example of a *recursive definition*. Note that if $m = 0$, (2.3) yields

$$y^1 = y^{0+1} = y^0 \cdot y = 1 \cdot y = y;$$

for $m = 1$,

$$y^2 = y^{1+1} = y^1 \cdot y = y \cdot y;$$

for $m = 2$,

$$y^3 = y^{2+1} = y^2 \cdot y = (y \cdot y) \cdot y;$$

for $m = 3$,

$$y^4 = y^{3+1} = y^3 \cdot y = ((y \cdot y) \cdot y) \cdot y.$$

Thus, the more sophisticated definition of powers is in agreement with the concept introduced in elementary books. In general, as in this example, the rather indefinite phrase "and so on" can be avoided by the use of recursive definitions.

Although the second definition has the air of greater precision, the question has really been begged. For, now the problem has become, "What justifies the assertion that for all natural numbers y and for all natural numbers k, y^k is a natural number?" Although we would accept, on the basis of (2.3) the claim that, say, 2^3 and 7^5 are natural numbers (indeed, using (2.3) and the theorems proved in this chapter, we could show easily that 2^3 and 7^5 are natural numbers) the proof that 872^{43792} is a natural number would present formidable difficulties. And even if this particular result, and others like it, could be obtained, we would be no closer to proving that for *all* natural numbers y and for *all* $k \in N$, y^k is a natural number.

The justification of the statement following (2.3), and others like it, require proofs which are tangential to the main purposes of this book. We shall omit most of these proofs but will discuss the nature of the problem at some length giving detailed examples to illustrate all the concepts introduced. It should be emphasized that the proofs required for all the theorems in the remainder of this chapter present no special difficulties. The interested reader will find detailed proofs of these theorems in Kerschner and Wilcox, *The Anatomy of Mathematics*, Chapter 11. The PFI plays an essential role in these proofs. The results, which are embodied in Theorems A through D and Theorem 65, below, will be used freely throughout Chapters 3, 4 and 5. We begin with

* Definition 27. For each $y \in N$, the function

$$m_y = \{(x, x \cdot y) \mid x \in N\}$$

is the *multiplication function for y*. $\mathfrak{M}$ is the set $\{m_y \mid y \in N\}$ of all multiplication functions.

In the customary notation, the multiplication function m_y is written

$$m_y(x) = x \cdot y, \; x \in N.$$

Now let us observe that if there is one and only one function $p_y : N \longrightarrow N$ satisfying the conditions:

$$\begin{cases} p_y(0) = 1, \\ p_y(n + 1) = m_y(p_y(n)), \; n \in N. \end{cases}$$

then

$$p_y(0) = 1,$$
$$p_y(1) = p_y(0 + 1) = m_y(p_y(0)) = 1 \cdot y = y$$
$$p_y(2) = p_y(1 + 1) = m_y(p_y(1)) = y \cdot y,$$
$$p_y(3) = p_y(2 + 1) = m_y(p_y(2)) = (y \cdot y) \cdot y,$$
$$p_y(4) = p_y(3 + 1) = m_y(p_y(3)) = ((y \cdot y) \cdot y) \cdot y.$$

Thus, the function p_y agrees with our intuitive ideas of what the powers of y ought to be. On the other hand, since p_y is a mapping of N into itself, it follows that for all $k \in N$, $p_y(k)$ is a natural number. Thus the entire problem of powers of natural numbers boils down to proving

* Theorem A. For each $y \in N$ there exists one and only one function $p_y : N \longrightarrow N$ such that

(2.4) $$\begin{cases} p_y(0) = 1, \\ p_y(n+1) = m_y(p_y(n)), \ n \in N.* \end{cases}$$

We accept Theorem A without proof.

*p_y is the *power function for* y and in accordance with customary notation we define the symbol "y^k" by

$$y^k = p_y(k),$$

for all $k \in N$. k is the *exponent*, y the *base*; y^k is the *k-th power of* y. Since the domain of p_y is N, it follows that p_y is a sequence. The sequence p_y, whose existence and uniqueness are assured by Theorem A, is said to be *defined recursively* by equations (2.4).*

EXAMPLE

Prove that $y^k y^l = y^{k+l}$ for all natural numbers k and l.

proof: By induction on l. For each $k \in N$ let

$$L_k = \{n \mid n \in N \text{ and } y^k y^n = y^{k+n}\}.$$

First, $0 \in L_k$ since $y^k \cdot y^0 = y^k \cdot 1 = y^k = y^{k+0}$. Next, L_k is hereditary. For, if $m \in L_k$, then

$$\begin{aligned}
y^k y^{m+1} &= y^k(y^m y^1) && \text{(Theorem A)} \\
&= (y^k y^m)y^1 && \text{(Associativity of } \cdot) \\
&= y^{k+m}y^1 && \text{(Since } m \in L_k) \\
&= y^{(k+m)+1} && \text{(Theorem A)} \\
&= y^{k+(m+1)} && \text{(Associativity of } +).
\end{aligned}$$

Hence, for each $k \in N$, L_k is hereditary, therefore $L_k = N$, and $y^k y^l = y^{k+l}$ holds for all natural numbers y, k and l.

EXERCISES

Prove by induction:

1. $(y^k)^l = y^{kl}$.

2. $y^k z^k = (yz)^k$.

The power functions p_y, $y \in N$, are special cases of a more general type of function, the so-called "generalized products." If a, $b \in N$, then the product ab is also a natural number. Since multiplication is a binary operation, a product of elements a, b, $c \in N$ can be defined only by reducing it to a successive performance of binary multiplications. Thus $a \cdot b \cdot c$ may be defined by

$$a \cdot b \cdot c = (a \cdot b) \cdot c.$$

Similarly, if a, b, c, $d \in N$, then the product $a \cdot b \cdot c \cdot d$ may be defined by

$$a \cdot b \cdot c \cdot d = ((a \cdot b) \cdot c) \cdot d.$$

At some point such as this, there is the usual temptation to say "and so on" and to regard a "generalized product"

(2.5) $a_1 \cdot a_2 \cdot \ldots \cdot a_k,\ a_i \in N,\ 1 \leq i \leq k,\ k \in N$

as being defined. The objections raised in the discussion of the intuitive concept of powers are pertinent here. Therefore we shall indicate what is involved in assigning precise meanings to expressions such as (2.5).

Let $\varphi : N_1 \longrightarrow N$ be a sequence of natural numbers. Then for each $n \in N_1$, $m_{\varphi(n)}$ is defined; in fact, it is the multiplication function

$$m_{\varphi(n)}(x) = x \cdot \varphi(n),\ x \in N.$$

Suppose, now, there exists one and only one function $\pi_\varphi : N \dashrightarrow N$ such that

$$\pi_\varphi(0) = 1,$$
$$\pi_\varphi(n+1) = m_{\varphi(n+1)}(\pi_\varphi(n)),\ n \in N.$$

To simplify the notation, set $\varphi(n) = a_n$, $n \in N$. If our assumptions concerning π_φ are justified, then for each $k \in N$, $\pi_\varphi(k)$ is a natural number. Moreover, if we give n the values 0, 1, 2, 3 in turn, then

$$\pi_\varphi(1) = m_{a_1}(\pi_\varphi(0)) = \pi_\varphi(0) \cdot a_1 = 1 \cdot a_1,$$
$$\pi_\varphi(2) = m_{a_2}(\pi_\varphi(1)) = \pi_\varphi(1) \cdot a_2 = a_1 \cdot a_2,$$
$$\pi_\varphi(3) = m_{a_3}(\pi_\varphi(2)) = \pi_\varphi(2) \cdot a_3 = (a_1 \cdot a_2) \cdot a_3,$$
$$\pi_\varphi(4) = m_{a_4}(\pi_\varphi(3)) = \pi_\varphi(3) \cdot a_4 = ((a_1 \cdot a_2) \cdot a_3) \cdot a_4,$$

and we see that the function π_φ agrees with the intuitive concept of a generalized product. Therefore, the problem of defining "generalized products" is reduced to proving

* Theorem B. Let $\varphi : N_1 \longrightarrow N$ be a sequence of natural numbers. Then there is one and only one function $\pi_\varphi : N \longrightarrow N$ such that

(2.6)
$$\begin{cases} \pi_\varphi(0) = 1, \\ \pi_\varphi(n + 1) = m_{\varphi(n+1)}(\pi_\varphi(n)), \ n \in N, \end{cases}$$

where, for each natural number k, $m_{\varphi(k)}$ is the multiplication function for $\varphi(k)$.*

We accept Theorem B without proof.

* Definition 28. π_φ is the *generalized product function for the sequence* φ.

For each $n \in N$ we define the symbol "$\prod\limits_{i=1}^{n} \varphi(i)$" by

$$\prod_{i=1}^{n} \varphi(i) = \pi_\varphi(n).$$

From the last part of Definition 28 it follows that

$$\prod_{i=1}^{0} \varphi(i) = \pi_\varphi(0) = 1$$

and

$$\prod_{i=1}^{n+1} \varphi(i) = \prod_{i=1}^{n} \varphi(i) \cdot \varphi(n + 1). \qquad \text{(Why?)}$$

Since the domain of π_φ is N, π_φ is a sequence. The sequence π_φ is said to be *defined recursively* by equations (2.6).*

EXERCISES

1. Let φ be the sequence $\varphi(n) = n$, $n \in N_1$. What is $\prod\limits_{i=1}^{5} \varphi(i)$, $\prod\limits_{i=1}^{6} \varphi(i)$? What are the common name and the common notation for $\prod\limits_{i=1}^{n} \varphi(i)$?

2. Prove Theorem A as a corollary of Theorem B. (Hint: Define φ by $\varphi(n) = y$ for all $n \in N_1$.) Hence, the power function is a special case of the generalized product function.

Next we consider the concept of "generalized sum." The problem here is quite analogous to the one involved in the definition of generalized product.

* Definition 29. For each $y \in N$, let s_y be the function

$$s_y(x) = x + y, \ x \in N;$$

thus $s_y = \{(x, x + y) \mid x \in N\}$. s_y is the *addition function for y*. We denote by "$\mathcal{S}$" the set $\{s_y \mid y \in N\}$ of all addition functions.

Theorem C. Let $\varphi : N \longrightarrow N$ be a sequence of natural numbers. Then there exists one and only one function $\sigma_\varphi : N \longrightarrow N$ such that

$$\textbf{(2.7)} \qquad \begin{cases} \sigma_\varphi(0) = 0 \\ \sigma_\varphi(n + 1) = s_{\varphi(n+1)}(\sigma_\varphi(n)), \ n \in N, \end{cases}$$

where, for each natural number k, $s_{\varphi(k)}$ is the addition function for $\varphi(k)$.*

Again we omit the proof but assume the validity of the theorem. For each $n \in N$, set $\varphi(n) = a_n$; then

$$\begin{aligned} \sigma_\varphi(1) &= s_{a_1}(\sigma_\varphi(0)) = \sigma_\varphi(0) + a_1 = 0 + a_1 = a_1, \\ \sigma_\varphi(2) &= s_{a_2}(\sigma_\varphi(1)) = \sigma_\varphi(1) + a_2 = a_1 + a_2, \\ \sigma_\varphi(3) &= s_{a_3}(\sigma_\varphi(2)) = \sigma_\varphi(2) + a_3 = (a_1 + a_2) + a_3, \\ \sigma_\varphi(4) &= s_{a_4}(\sigma_\varphi(3)) = \sigma_\varphi(3) + a_4 = ((a_1 + a_2) + a_3) + a_4, \end{aligned}$$

so that the function σ_φ agrees with the intuitive notion of a "generalized sum."

As is the case with the earlier generalized operations, σ_φ is a sequence since its domain is N.

* Definition 30. σ_φ is the *generalized sum function* for the sequence φ; it is said to be *defined recursively* by equations (2.7). Further, the symbol "$\sum\limits_{i=1}^{n} \varphi(i)$" is defined by

$$\sum_{i=1}^{n} \varphi(i) = \sigma_\varphi(n).\text{*}$$

From the last part of Definition 30, it follows that

$$\sum_{i=1}^{0} \varphi(i) = 0$$

and

$$\sum_{i=1}^{n+1} \varphi(i) = \sum_{i=1}^{n} \varphi(i) + \varphi(n + 1). \qquad \text{(Why?)}$$

If φ is the sequence $\varphi(n) = a$, $n \in N$, then

$$\sum_{i=1}^{n} \varphi(i) = \underbrace{a + a + \ldots + a.}_{n \text{ times}}$$

EXERCISES

Prove by induction:

1. $\underbrace{a + a + \ldots + a}_{n \text{ times}} = n \cdot a$; i.e., if $\varphi(n) = a$, $n \in N$, then

$$\sum_{i=1}^{n} \varphi(i) = n \cdot a.$$

2. Suppose $a \in N$ and suppose there is a natural number b such that $a = xb$, $x \in N_1$, then we define "a/x" by

$$\frac{a}{x} = b.$$

Prove: For all $n \in N$, $n(n+1)/2 \in N$.

3. If $\varphi(n) = n$, $n \in N$, then $\sum_{i=1}^{n} \varphi(i) = n(n+1)/2$.

4. Prove: If $z \in N$, $z > 1$, then $(z^n - 1)/(z - 1) \in N$ for all $n \in N$.

5. Prove: For all $a, z, n \in N$, $z > 1$,

$$a + az + az^2 + \ldots + az^n = \frac{a(z^{n+1} - 1)}{z - 1}$$

The reader has no doubt observed that Theorems A, B and C resemble each other very closely; the only difference among them lies in the nature of the functions m_y and s_y. From the resemblances, one might infer that all three theorems are special cases of a more general result. That is true, and we now state (without proof) a theorem of which A, B and C are corollaries.

* Theorem D. Let $\mathfrak{F}$ be a set of functions such that for all $f \in \mathfrak{F}$, $\mathfrak{D}(f) = N$ and $\mathfrak{R}(f) \subset N$, and let φ be a sequence of elements of $\mathfrak{F}$. Let $\varphi(n) = \varphi_n$, $n \in N$. Further, let $a \in N$. Then there exists one and only one function $\Gamma : N \longrightarrow N$ such that

(2.8)
$$\begin{cases} \Gamma(0) = a, \\ \Gamma(n + 1) = \varphi_{n+1}(\Gamma(n)), \ n \in N.* \end{cases}$$

If $\mathfrak{F} = \mathfrak{M}$, the set of multiplication functions, then Theorem B results; if $\mathfrak{F} = \mathfrak{S}$, the set of addition functions, then we obtain Theorem C.   The sequence Γ is said to be *defined recursively* by equations (2.8).*

Although Theorem D is adequate for the elementary (school) arithmetic of the natural numbers, it is not sufficiently inclusive to yield all the generalized operations that we shall need.   Therefore we state (again, without proof) one more theorem of which D is a special case, and from which all our generalized operations can be obtained.

* Theorem 62.   Let A be a set and let $\mathfrak{K}$ be a set of functions such that for all $f \in \mathfrak{K}$, $\mathfrak{D}(f) = A$ and $\mathfrak{R}(f) \subset A$.   Further, let φ be a sequence of functions in $\mathfrak{K}$, where we set $\varphi(n) = \varphi_n$, $n \in N_1$.   Then for each $a \in A$, there exists one and only one function $\Lambda : N \longrightarrow A$ such that

(2.9)
$$\begin{cases} \Lambda(0) = a, \\ \Lambda(n + 1) = \varphi_{n+1}(\Lambda(n)), \ n \in N. \end{cases}$$

The sequence Λ is said to be *defined recursively* by equations (2.9).*

EXERCISES

1.   Construct some examples of sequences Λ in accordance with Theorem 62.

2.   Prove that Theorem D is a corollary of Theorem 62.

In order that our arithmetic theory may have greater flexibility, other generalized results are needed.   For instance, the associative law for the addition of natural numbers states:

For all a, b, $c \in N$,

$$(a + b) + c = a + (b + c).$$

Using the associative law repeatedly, it is possible to prove that for all a, b, c, $d \in N$,

$$\begin{aligned} ((a + b) + c) + d &= (a + b) + (c + d) = a + (b + (c + d)) \\ &= (a + (b + c)) + d = a + ((b + c) + d). \end{aligned}$$

But, for even as few as five elements a, b, c, d, $e \in N$, the proof that

$$(((a + b) + c) + d) + e = (((a + (b + c)) + d) + e$$
$$= (a + (b + c)) + (d + e)$$
$$= \text{etc.}$$

is quite tedious. And there is still the matter of "and so on" to be taken care of. Clearly, a "generalized associative law" for addition, which permits the insertion of parentheses in any meaningful way, is desirable. Such a law can be stated and proved for any set S on which is defined an associative binary operation. From this theorem, it follows that "generalized associative laws" hold both for addition and for multiplication of natural numbers, as well as for the operations in the other systems that will be studied later. Further, if the binary operation under consideration is commutative, then we can state and prove a "generalized commutative law." To go into these matters would require a lengthy digression from our main argument. We assume the results (in an intuitive way) and refer the reader to the book by Kerschner and Wilcox.

EXERCISES

Prove by induction:

1. $a \cdot (b_1 + b_2 + \ldots + b_n) = a \cdot b_1 + a \cdot b_2 + \ldots + a \cdot b_n$ for all natural numbers $a, b_1, b_2, \ldots, b_n$ and for all $n \in N$.

2. $x^n - y^n = (x - y) \cdot (x^{n-1} + x^{n-2}y + \ldots + xy^{n-2} + y^{n-1})$ for all natural numbers x, y and for all $n \in N_1$.

3. $a_1^2 + a_2^2 + \ldots + a_n^2 \geq 0$ for all natural numbers $a_1, a_2, \ldots, a_n$. Show that $a_1^2 + a_2^2 + \ldots + a_n^2 = 0$ if and only if $a_1 = a_2 = \ldots = a_n = 0$.

We conclude this section (and Chapter 2) with some examples of the misuse of the PFI.

EXAMPLES

1. For all natural numbers $n \geq 1$, $1 + 2 + 3 + \ldots + n = (n^2 + n + 1)/2$.
   "*Proof:*" Assume the equation holds for $k \in N_1$; in other words, assume that

$$1 + 2 + 3 + \ldots + k = \frac{k^2 + k + 1}{2}.$$

Then

$$1 + 2 + 3 + \ldots + k + (k + 1) = \frac{k^2 + k + 1}{2} + (k + 1).$$

But

$$\frac{k^2 + k + 1}{2} + (k + 1) = \frac{k^2 + k + 1 + 2(k + 1)}{2}$$

$$= \frac{(k + 1)^2 + (k + 1) + 1}{2},$$

therefore the given equation holds for all $n \in N_1$.

2. Although the real numbers have not yet been constructed (Chapter 4) nor have polynomial functions been defined (Book II), we take the liberty of assuming a small amount of knowledge of these things for the purposes of this example.

Recall, first, that a *zero* of a polynomial function

$$f(x) = a_0 x^n + a_1 x^{n-1} + \ldots + a_{n-1} x + a_n, \ a_0 \neq 0, \ x \in \text{set}$$

of real numbers, $a_0, a_1, \ldots, a_n$ real, is a number r such that

$$f(r) = 0.$$

Next, consider the polynomial function $g(x) = x^2 - 2x + 1$, x real. Clearly, one is a zero of f since $g(1) = 0$. But we call one a zero of g of *multiplicity two* because $g(x) = (x - 1)^2$, x real. Similarly, one is a zero of

$$h(x) = (x - 1)^3 \cdot (x + 2), \ x \text{ real},$$

of *multiplicity three*, -2 is a zero of h of *multiplicity one*.

Definition. A number r is a zero of a polynomial f of *multiplicity k*, $k \in N$, if and only if there is a polynomial g such that

1. $f(x) = (x - r)^k \cdot g(x)$, x real; and
2. $x - r$ is not a factor of g.

Now, a well-known theorem of algebra states:

Let $n \in N$. For all polynomials of degree $n > 0$ (with real coefficients), the sum of the multiplicities of the real zeros of f is at most n.

We "prove" the following strange form of the theorem:

Let $n \in N$. For all polynomials of degree $n \geq 1$ (with real coefficients), the sum of the multiplicities of the real zeros of f is exactly n.

"proof:" Clearly, every polynomial function of degree one (with real coefficients) has exactly one real zero. Namely, for all real numbers a, b, $a \neq 0$, the zero of $ax + b$, x real, is $-b/a$.

Let $k \in N$ and assume the result is true for all polynomial functions (with real coefficients) of degree k, and let

$$f(x) = a_0 x^{k+1} + a_1 x^k + \ldots + a_{k+1}, \ x \text{ real,}$$

where $a_0, a_1, \ldots, a_{k+1}$ are real and $a_0 \neq 0$. If r is a real zero of f, then

$$f(r) = a_0 r^{k+1} + a_1 r^k + \ldots + a_{k+1} = 0,$$

whence, for all x,

$$\begin{aligned} f(x) &= f(x) - f(r) \\ &= a_0(x^{k+1} - r^{k+1}) + a_1(x^k - r^k) + \ldots + a_k(x - r). \end{aligned}$$

Now, by an argument similar to that required for Exercise 2, page 128, it can be proved that for all real numbers x, r and for all natural numbers m,

$$x^m - r^m = (x - r)(x^{m-1} + x^{m-2}r + \ldots + xr^{m-2} + r^{m-1}).$$

Hence,

$$\begin{aligned} f(x) = f(x) - f(r) &= a_0(x - r)(x^k + x^{k-1}r + \ldots + r^k) \\ &\quad + a_1(x - r)(x^{k-1} + x^{k-2}r + \ldots + r^{k-1}) \\ &\quad + \ldots + a_k(x - r) \\ &= (x - r)\{a_0(x^k + x^{k-1}r + \ldots + r^k) \\ &\quad + a_1(x^{k-1} + x^{k-2}r + \ldots + r^{k-1}) + \ldots + a_k\} \\ &= (x - r) \cdot (a_0 x^k + b_1 x^{k-1} + \ldots + b_k), \end{aligned}$$

for all real x, where $a_0, b_1, \ldots, b_k$ are real numbers. The polynomial function $x - r$, x real, has the real zero r of multiplicity one, and for the polynomial function

$$a_0 x^k + b_1 x^{k-1} + \ldots + b_k, \ x \text{ real,}$$

the sum of the multiplicities of the real zeros is, by assumption, equal to k. Hence, the sum of the multiplicities of the real zeros of

$$(x - r)(a_0 x^k + a_1 x^{k-1} + \ldots + a_k)$$

is exactly $k + 1$.   Hence, the assertion holds for all natural numbers $n \geq 1$.

Exercise

What is wrong with the arguments in the above two examples?

3

THE INTEGERS AND THE
RATIONAL NUMBERS

3.1. INTRODUCTION

The numbers we have constructed so far are clearly inadequate for many purposes. For example, if we wished, we could introduce the operation of subtraction in the set of natural numbers by means of the definition

$m - n$ is the natural number x, if there is any, such that $m = n + x$.

The qualification "if there is any" is essential; for, if $n > m$, there is no such natural number. Thus, unrestricted subtraction cannot be carried out in the set of natural numbers.

Similarly, the operation of division could be defined in the set of natural numbers by means of

for $n \neq 0$, $m \div n$ is the natural number x, if there is any, such that $n \cdot x = m$.

Again the qualification is essential, since obviously there are many instances in which such a natural number will not exist for given natural numbers $n \neq 0$ and m.

In order to extend the scope of our arithmetic, we shall, in the present chapter, construct two new systems of numbers, the integers and the rationals.   The system of integers will admit unrestricted subtractions, and we shall prove that the integers contain a "copy" of the natural numbers.   Similarly, the system of rational numbers will admit unrestricted division, except by zero, as well as unrestricted subtraction, and it will be proved that the system of rational numbers contains a "copy" of the integers.   Thus, the set of integers will constitute an "extension" of the natural numbers, and the rational numbers will be an "extension" of the integers.

3.2.   DEFINITION AND PROPERTIES OF THE INTEGERS

We wish to define the integers in such a way that not only can subtraction (as well as addition and multiplication) be carried out for all pairs of integers, but also so that, for example, $2 - 7$, $5 - 10$, etc., can be identified as the integer we shall eventually denote by "-5."   To achieve these objectives, we begin by defining a relation $\sim$ on $N \times N$.

* Definition 1.   For all (a,b), $(c,d) \in N \times N$, $(a,b) \sim (c,d)$ if and only if $a + d = c + b$.   Alternatively, this definition can be stated:

> $\sim$ is the subset of $(N \times N) \times (N \times N)$ of those elements $((a,b), (c,d))$, such that $a + d = c + b$.*

We have in mind that ultimately the symbol "(a,b)" will denote $a - b$, and that $(a,b) \sim (c,d)$ will mean that $a - b = c - d$; this is true, of course, if and only if $a + d = c + b$.   Clearly, this is in accord with our experiences with elementary arithmetic.

*Theorem 1. $\sim$ is an equivalence relation on $N \times N$.

proof: (1) For all $(a,b) \in N \times N$ it is clear that $(a,b) \sim (a,b)$; hence $\sim$ is reflexive.

(2) For all (a,b), $(c,d) \in N \times N$, if $(a,b) \sim (c,d)$, then $a + d = c + b$, whence $c + b = a + d$ and therefore $(c,d) \sim (a,b)$.   Thus $\sim$ is symmetric.

(3) Finally, suppose that $(a,b) \sim (c,d)$ and $(c,d) \sim (e,f)$.   We must prove $(a,b) \sim (e,f)$.   Since $(a,b) \sim (c,d)$ we have $a + d = c + b$; hence

$a + d + f = c + b + f$. Also, since $(c,d) \sim (e,f)$, we have $c + f = e + d$; hence $c + f + b = e + d + b$. Since $c + b + f = c + f + b$, we deduce $a + d + f = e + d + b$. By the cancellation law for addition in N, it follows that $a + f = e + b$; that is, $(a,b) \sim (e,f)$.

q.e.d.

By Theorem 12 of Chapter 1 we know that $\sim$ determines a partition of $N \times N$; this partition will be denoted by "Z". The elements of Z are subsets $[a,b]$ of $N \times N$ defined by

$$[a,b] = \{(x,y) \mid (x,y) \sim (a,b)\}.$$

Definition 2. The elements $[a,b]$ in Z are *integers*; Z is the *set of integers*.

Theorem 2. $[a,b] = [c,d]$ if and only if $a + d = c + b$.

proof: Suppose $[a,b] = [c,d]$; since $(a,b) \in [a,b]$, it follows that $(a,b) \in [c,d]$. Hence, by definition of $[c,d]$, $(a,b) \sim (c,d)$ and therefore, by Definition 1, $a + d = c + b$.

Conversely, if $a + d = c + b$ then (Definition 1) $(a,b) \sim (c,d)$ and $(a,b) \in [c,d]$. But $(a,b) \in [a,b]$. Thus $[a,b]$ and $[c,d]$ are two elements of a partition (namely, Z) having (a,b) in common. By Definition 29, Chapter 1, $[a,b] = [c,d]$.

q.e.d.*

Our intuitive knowledge of elementary arithmetic suggests that the sum of, say, $4 - 1$ and $5 - 3$ might be defined as $9 - 4 = (4 + 5) - (1 + 3)$. However, such a definition does not have quite the desired effect since $4 - 1$ and $5 - 3$ are not integers. An integer, as we have defined it, is a set of "differences" [1] so that the integer "3" [1] consists of all the "differences" $4 - 1, 5 - 2, 6 - 3, \ldots$, of natural numbers. Thus we shall want to define sums, not of ordered pairs of integers, but rather of sets of ordered pairs of integers. But here a subtle point arises: since $[3,0] = [4,1] = [5,2] = $ etc., our definition must be so constructed that in any sum involving the set $[3,0]$, $[3,0]$ may be replaced by $[4,1]$, etc., without altering the sum.

In introducing the concept of product, we are led to similar considerations. The definition of the product of $4 - 1$ and $5 - 3$ might be taken

---

[1] We have put the word "differences" and the symbol "3" in quotes because the word "difference" has not yet been defined and because we have not yet introduced "3" as the symbol for the set $[3,0]$.

as $23 - 17 = 4 \cdot 5 + 1 \cdot 3 - (1 \cdot 5 + 4 \cdot 3)$, but objections similar to those noted in connection with the definition of sum must be taken into account.   Before giving our definition of sum and product of integers, we require a theorem which will enable us to state the desired definitions of sums and products of sets $[a,b]$ and $[c,d]$.

* Theorem 3.   If $[a,b] = [c,d]$ and $[e,f] = [g,h]$ then

(i) $[a + e, b + f] = [c + g, d + h]$, and
(ii) $[ae + bf, af + be] = [cg + dh, ch + dg]$.

proof: (i) By Theorem 2, from $[a,b] = [c,d]$ and $[e,f] = [g,h]$ follow $a + d = c + b$ and $e + h = g + f$, respectively, and therefore, $(a + d) + (e + h) = (c + b) + (g + f)$.   By the commutative and associative laws for the addition of natural numbers

$$(a + e) + (d + h) = (c + g) + (b + f);$$

and again, by Theorem 2, this means that

$$[a + e, b + f] = [c + g, d + h].$$

(ii) This part of the proof is carried out easily in two steps.   We show first that

$$[ae + bf, af + be] = [ce + df, cf + de],$$

or, by Theorem 2, that

$$(ae + bf) + (cf + de) = (ce + df) + (af + be).$$

Now,

$$
\begin{aligned}
(ae + bf) + (cf + de) & \\
= (a + d)e + (b + c)f & \quad \text{(Com., assoc., dist. laws} \\
& \quad \text{for natural numbers)} \\
= (b + c)e + (a + d)f & \quad \text{(Since } a + d = b + c) \\
= (ce + df) + (af + be). & \quad \text{(Com., assoc., dist. laws} \\
& \quad \text{for natural numbers)}
\end{aligned}
$$

Next we show that

$$[ce + df, cf + de] = [cg + dh, ch + dg]$$

or

$$(ce + df) + (ch + dg) = (cg + dh) + (cf + de).$$

Using $e + h = g + f$, and proceeding as above, the last equation is established easily. Since equality is transitive, part (ii) of the theorem is proved.

$$\text{q.e.d.}_*$$

The next step in the definitions of sum and product of integers is a corollary to Theorem 3, which requires a complete understanding of the definition of binary operation. We therefore urge the reader to review carefully Definition 27 of Chapter 1.

corollary: The sets

$$\oplus = \big\{(([a,b], [e,f]), [a + e, b + f]) \mid [a,b], [e,f] \in Z\big\}$$

and

$$\odot = \big\{(([a,b], [e,f]), [ae + bf, af + be]) \mid [a,b], [e,f] \in Z\big\}$$

are binary operations on Z. In other words, both $\oplus$ and $\odot$ are mappings of $Z \times Z$ into Z.

proof: Consider first the set $\oplus$. Observe that $\oplus$ is a set of ordered pairs and the left member of each ordered pair

(3.1)
$$([a,b], [e,f])$$

is, in turn, an ordered pair of integers. The right member, $[a + e, b + f]$, of the ordered pair having $([a,b], [e,f])$ as left member, is again an integer. Thus $\oplus$ is a relation between the sets $Z \times Z$ and Z. Further, by definition of $\oplus$, $\mathfrak{D}(\oplus) = Z \times Z$, and $\mathfrak{R}(\oplus) \subset Z$. Therefore, in order to prove that $\oplus$ is a binary operation, we show (Definition 27, Chapter 1) that if

(3.2)
$$([a,b], [e,f]) = ([c,d], [g,h]),$$

then

(3.3)
$$[a + e, b + f] = [c + g, d + h].$$

But, from (3.2) we deduce that $[a,b] = [c,d]$ and $[e,f] = [g,h]$; therefore by Theorem 3, $[a + e, b + f] = [e + g, d + h]$, i.e., (3.3) holds and $\oplus$ is a binary operation on Z.

In a similar fashion, $\odot$ is a set of ordered pairs where left members are given by (3.1) and corresponding right members are denoted by

$$[ae + bf, af + be].$$

Therefore $\odot$ is a relation between $Z \times Z$ and Z. Clearly $\mathfrak{D}(\odot) =$

$Z \times Z$ and $\mathcal{R}(\odot) \subset Z$. To prove that $\odot$ is a binary operation on Z we need only show that if (3.2) holds, then

(3.4) $[ae + bf,\, af + be] = [cg + dh,\, ch + dg].$

As before, (3.2) yields $[a,b] = [c,d]$, $[e,f] = [g,h]$, whence by Theorem 3 we deduce equation (3.4).   Hence, $\odot$ is also a binary operation on Z.

q.e.d.

Definition 3.   The binary operation $\oplus$ is the *sum* of integers and the binary operation $\odot$ is the *product* of integers.

With the customary notation for mappings, $\oplus$ and $\odot$ are written

$$\oplus\, (([a,b],\, [e,f])) = [a + e,\, b + f], \text{ for all } [a,b],\, [e,f] \in Z$$

and

$$\odot\, (([a,b],\, [e,f])) = [ae + bf,\, af + be], \text{ for all } [a,b],\, [e,f] \in Z,$$

respectively; and with the preferred notation for binary operations these become

(3.5) $[a,b] \oplus [e,f] = [a + e,\, b + f], \quad [a,b],\, [e,f] \in Z$

and

(3.6) $[a,b] \odot [e,f] = [ae + bf,\, af + be], \quad [a,b],\, [e,f] \in Z,$

respectively.*

It must be emphasized that the symbols "$\oplus$" and "$+$" appearing in equation (3.5) do not have the same meaning.   The latter, "$+$," is the symbol for the sum of natural numbers; it was defined in Chapter 2 and is now being used to define the new symbol, "$\oplus$," which denotes the sum of integers.   Similarly, the symbols "$\odot$" and "$\cdot$" appearing in equation (3.6) have different meanings.

The examples below show that the definitions of $\oplus$ and $\odot$ agree with our intuitive notions of sums and products of integers.

EXAMPLES

1. $[4,1] \oplus [5,3] = [4 + 5,\, 1 + 3] = [9,4]$.   If we recall that $[a,b]$ is to be interpreted as the set of all differences which are equal to $a - b$, then this example simply illustrates that, for integers, $3 \oplus 2 = 5$.

2. $[4,1] \odot [5,3] = [23,17]$, so that this is the statement, for integers, that $3 \odot 2 = 6$.

3. $[1,4] \oplus [5,3] = [6,7]$ and this agrees with $(-3) \oplus 2 = -1$.

4. $[1,4] \odot [5,3] = [17,23]$, or $(-3) \odot 2 = -6$.

5. $[1,4] \odot [3,5] = [23,17]$, i.e., $(-3) \odot (-2) = 6$.

6. $[2,4] \oplus [5,4] = [7,8]$, and this agrees with $(-2) + 1 = -1$.

7. $[2,4] \odot [5,4] = [26,28]$, which agrees with $(-2) \cdot 1 = -2$.

The following theorem will be useful in simplifying the notations and later proofs.

* Theorem 4. (i) $[0,0]$ is a "zero element" for Z; namely, for all integers $[a,b]$, $[a,b] \oplus [0,0] = [a,b]$ and $[a,b] \odot [0,0] = [0,0]$. $[a,b] = [0,0]$ if and only if $a = b$.

(ii) For all integers $[a,b]$, if $a > b$, there is a unique natural number x such that $[a,b] = [x,0]$.

(iii) For all integers $[a,b]$, if $a < b$, there is a unique natural number y such that $[a,b] = [0,y]$.

proof: (i) Clearly, $[0,0] = [x,x]$ for all $x \in N$, by Theorem 2. The remaining statements follow from Definition 3.

(ii) If $a > b$ there is a unique natural number x such that $a = x + b$. Then $a + 0 = b + x$ and, by Theorem 2, $[a,b] = [x,0]$. If x and x_1 are such that $[a,b] = [x,0]$ and $[a,b] = [x_1,0]$, then $[x,0] = [x_1,0]$, so $x + 0 = x_1 + 0$, or $x = x_1$.

(iii) The proof of (iii) parallels that of (ii).*

By reason of Theorem 4, each integer can be represented by exactly one of "$[0,0]$," "$[x,0]$," "$[0,x]$," where x is a nonzero natural number.

* Definition 4. If $[a,b]$, $[c,d]$ and $[e,f]$ are integers, then

$$[a,b] \oplus [c,d] \oplus [e,f] = ([a,b] \oplus [c,d]) \oplus [e,f]$$

and

$$[a,b] \odot [c,d] \odot [e,f] = ([a,b] \odot [c,d]) \odot [e,f].$$

Theorem 5. (i) Addition and multiplication of integers are commutative and associative; (ii) multiplication is distributive over addition; (iii) the cancellation laws for addition and for multiplication hold in Z.

proof: (i) Exercise.

(ii) $[a,b] \odot ([c,d] + [e,f])$

$$= [a,b] \odot [c + e, d + f]$$
$$= [a(c + e) + b(d + f), a(d + f) + b(c + e)]$$
$$= [(ac + bd) + (ae + bf), (ad + bc) + (af + be)]$$
$$= [ac + bd, ad + bc] \oplus [ae + bf, af + be]$$
$$= [a,b] \odot [c,d] \oplus [a,b] \odot [e,f].$$

(iii) The cancellation law for addition in Z.   We prove:

If $[a,b] \oplus [c,d] = [a,b] \oplus [e,f]$, then $[c,d] = [e,f]$.   The hypothesis implies that

$$[a + c, b + d] = [a + e, b + f]$$

whence

$$(a + c) + (b + f) = (a + e) + (b + d).$$

By the associative and commutative laws for addition in N,

$$(a + b) + (c + f) = (a + b) + (e + d).$$

Since the cancellation law for addition holds in N, we have

$$c + f = e + d$$

whence, by Theorem 2,

$$[c,d] = [e,f].$$

The cancellation law for multiplication in Z.   We prove: If $[a,b] \odot [c,d] = [a,b] \odot [e,f]$ and $[a,b] \neq [0,0]$, then $[c,d] = [e,f]$.

Suppose, first, $[a,b] = [x,0]$, $x \neq 0$.   Then

$$[x,0] \odot [c,d] = [x,0] \odot [e,f],$$

or

$$[xc,xd] = [xe,xf],$$

whence

$$xc + xf = xe + xd,$$

i.e.,

$$x(c + f) = x(e + d).$$

Since $x \neq 0$ and, since the cancellation law holds for multiplication in N,

$$c + f = e + d,$$

whence

$$[c,d] = [e,f].$$

If $[a,b] = [0,x]$, $x \neq 0$, a similar proof shows that $[c,d] = [e,f]$. This completes the proof of Theorem 5.∗

EXERCISE

Prove: $[1,0] \odot [a,b] = [a,b] \odot [1,0] = [a,b]$.

Before we introduce the concept of order for the integers, we require a simple result which is left to the reader to prove:

* If $[a,b] = [a_1,b_1]$ and $[c,d] = [c_1,d_1]$ and if $a + d < c + b$, then $a_1 + d_1 < c_1 + b_1$.

Definition 5. $[a,b]$ is *less than* $[c,d]$ if and only if $a + d < c + b$. If $[a,b]$ is less than $[c,d]$ we write

$$[a,b] <' [c,d].$$

$[a,b]$ is *greater than* $[c,d]$ if and only if $[c,d]$ is less than $[a,b]$. If $[a,b]$ is greater than $[c,d]$ we write

$$[a,b] >' [c,d].$$

$<'$ and $>'$ are *strict inequalities*. $[a,b] \leq' [c,d]$ means that $[a,b] <' [c,d]$ or $[a,b] = [c,d]$. Similarly, $[a,b] \geq' [c,d]$ means that $[a,b] >' [c,d]$ or $[a,b] = [c,d]$. $\leq'$ and $\geq'$ are *weak inequalities*.*

EXERCISE

Define (a) $<'$, (b) $>'$, (c) $\leq'$, (d) $\geq'$ as subsets of $Z \times Z$.

The symbol " $<'$ " does not denote the same relation as the symbol " $<$ ". The latter denotes a relation on the set N of natural numbers and is used to define the former, which denotes a relation on the set of integers. Also, we note that Definition 5 is consistent with experience. For example, according to this definition $[2,5] >' [3,9]$ since $2 + 9 > 3 + 5$. Keeping in mind the eventual interpretation of $[a,b]$ in terms of $a - b$, we see that our definition leads to a familiar result.

* **Theorem 6.** If $[a,b]$, $[c,d]$ are integers, then exactly one of

$$[a,b] <' [c,d], \quad [a,b] = [c,d], \quad [c,d] <' [a,b]$$

holds. Further, if $[a,b] <' [c,d]$ and $[c,d] <' [e,f]$ then $[a,b] <' [e,f]$.

proof: Since exactly one of

$$a + d < c + b, \quad a + d = c + b, \quad c + b < a + d$$

holds for all natural numbers a, b, c, d, the first statement is an immediate consequence of Definition 5 and Theorem 2.

Now, suppose $[a,b] <' [c,d]$ and $[c,d] <' [e,f]$. Then $a + d < c + b$ and $c + f < e + d$. From the order properties for the natural numbers, we deduce

$$(a + d) + f < (c + b) + f,$$

or

$$(a + f) + d < (c + f) + b.$$

Similarly, we obtain $(c + f) + b < (e + d) + b$, or $(c + f) + b < (e + b) + d$. Therefore

$$(a + f) + d < (e + b) + d,$$

whence

$$a + f < e + b,$$

and consequently,

$$[a,b] <' [e,f].$$

q.e.d.*

EXERCISES

Prove:

1. If $[a,b] >' [c,d]$ and $[c,d] >' [e,f]$ then $[a,b] >' [e,f]$.

2. If $[a,b] >' [c,d]$ and $[c,d] \geq' [e,f]$ then $[a,b] >' [e,f]$.

3. If $[a,b] \geq' [c,d]$ and $[c,d] >' [e,f]$ then $[a,b] >' [e,f]$.

4. If $[a,b] \geq' [c,d]$ and $[c,d] \geq' [e,f]$ then $[a,b] \geq' [e,f]$.

5. If $[a,b] >' [c,d]$ then $[a,b] \oplus [e,f] >' [c,d] \oplus [e,f]$.   Prove a similar theorem for the weak inequality.

6. If $[a,b] \oplus [e,f] >' [c,d] \oplus [e,f]$ then $[a,b] >' [c,d]$; a similar result holds for the weak inequality.

* Definition 6. If $[a,b] >' [0,0]$ then $[a,b]$ is *positive*; if $[a,b] <' [0,0]$ then $[a,b]$ is *negative*.

$[a,b] >' [0,0]$ means $a = a + 0 > 0 + b = b$.   But $a > b$ means that there is one and only one natural number $x \neq 0$ such that $a = b + x$; hence $[a,b] = [b + x, b] = [x,0]$ for exactly one natural number $x \neq 0$. Likewise, if $[a,b] <' [0,0]$, then $[a,b] = [0,y]$ for exactly one natural number $y \neq 0$. By Theorem 6, every integer is either positive or negative or $[0,0]$.

Theorem 7. If $[a,b] \geq' [c,d]$ and $[e,f]$ is positive, then

$$[a,b] \odot [e,f] \geq' [c,d] \odot [e,f].$$

If $[e,f]$ is negative, then $[a,b] \odot [e,f] \leq' [c,d] \odot [e,f]$. Similar results hold for the strict inequalities.

proof: Exercise.∗

We now introduce subtraction.

∗ Theorem 8. For each integer $[a,b]$ there is a unique integer $[x,y]$ such that $[a,b] \oplus [x,y] = [0,0]$.

proof: $[a,b] \oplus [b,a] = [a + b, b + a] = [a + b, a + b] = [0,0]$; hence the required integer exists. Its uniqueness follows from the cancellation law for addition of integers.

q.e.d.

Definition 7. The integer $[b,a]$ is the (*additive*) *inverse* of $[a,b]$. It is also denoted by

$$-[a,b].$$

Definition 8. The integer $[a + d, b + c]$ is the *difference of* $[a,b]$ *and* $[c,d]$; it is also denoted by

$$[a,b] - [c,d].$$

The process of computing a difference is *subtraction*.

It is an easy task (for the reader) to prove: If $[a,b] = [a_1,b_1]$ and $[c,d] = [c_1,d_1]$, then $[a,b] - [c,d] = [a_1,b_1] - [c_1,d_1]$.

Theorem 9. For every pair of integers $[a,b]$, $[c,d]$ there is a unique integer $[x,y]$ such that $[a,b] = [c,d] \oplus [x,y]$. $[x,y]$ is the difference of $[a,b]$ and $[c,d]$.

proof: The integer $[a + d, b + c] = [a,b] \oplus [d,c]$ has the desired property, since

$$[c,d] \oplus [a + d, b + c] = [c + (a + d), d + (b + c)]$$
$$= [a + (c + d), b + (c + d)]$$
$$= [a,b] \oplus [c + d, c + d]$$
$$= [a,b].$$

Again, the cancellation law for addition shows that the integer $[a + d, b + c]$ is unique.

q.e.d.*

EXERCISES

1. Prove that $-(-[a,b]) = [a,b]$.

2. Prove: $[a,b] \oplus (-[c,d]) = [a,b] - [c,d]$.

3. Prove: $[x,x] \oplus (-[a,b]) = [b,a]$.

Our next step is to show that the integers contain a "copy" of the natural numbers. Our object is to show that the set Z contains a subset $\mathfrak{N}$ whose elements behave arithmetically exactly as do the natural numbers. In saying that $\mathfrak{N}$ and N "behave" alike arithmetically, we mean that they behave alike as far as addition, multiplication and order are concerned. Each of the sets $\mathfrak{N}$ and N respectively may have properties which are not shared by the other. This, however, is of no concern to us since it is only arithmetic properties that we wish to study. To make these ideas precise, we introduce the concept of "isomorphism."

* Definition 9. A subset A of Z is *isomorphic* with N if and only if there is a one-one correspondence f between N and A such that:

1. $f(x + y) = f(x) \oplus f(y)$;
2. $f(x \cdot y) = f(x) \odot f(y)$;
3. If $x > y$ then $f(x) >' f(y)$.

The correspondence f is an *isomorphism* between N and A.[2]

Theorem 10. The subset $\mathfrak{N} = \{[x,0] \mid x \in N\}$, of Z, is isomorphic with N.

---

[2] The present definition of isomorphism is quite limited in its applications. The concept enters into many branches of mathematics and more general definitions can be given so as to meet a variety of needs. For our present purposes, the above definition suffices.

ꝉroof: In order to prove the theorem, we exhibit a one-one correspondence f between N and $\mathfrak{N}$ having the three properties required in the definition. Define f by: for all $x \in N$,

$$f(x) = [x,0].$$

Obviously, $\mathfrak{D}(f) = N$ and $\mathfrak{R}(f) = \mathfrak{N}$ so that the mapping f is onto. Moreover, since $[x,0] = [y,0]$ if and only if $x = y$, the correspondence is one-one.

1. $f(x + y) = [x + y, 0] = [x,0] \oplus [y,0] = f(x) \oplus f(y)$.
2. $f(x \cdot y) = [x \cdot y, 0] = [x,0] \odot [y,0] = f(x) \odot f(y)$.
3. $[x,0] >' [y,0]$ if and only if $x = x + 0 > y + 0 = y$, and therefore the third property holds.

q.e.d.∗

EXERCISE

Is the correspondence g, defined by $g(x) = [0,x]$, an isomorphism between N and the set $\{[0,x] \mid x \in N\}$? Why?

∗Theorem 11. The Principle of Finite Induction holds for the set $\mathfrak{N}$ of nonnegative integers. More precisely: If $\mathfrak{M}$ is a set of nonnegative integers such that $[0,0] \in \mathfrak{M}$ and $[n + 1, 0] \in \mathfrak{M}$ whenever $[n,0] \in \mathfrak{M}$, then $\mathfrak{M} = \mathfrak{N}$.

proof: The hypotheses of the theorem are:

1. $\mathfrak{M} \subset \mathfrak{N}$,
2. $[0,0] \in \mathfrak{M}$, and
3. If $[n,0] \in \mathfrak{M}$ then $[n + 1, 0] \in \mathfrak{M}$.

We wish to prove that $\mathfrak{M} = \mathfrak{N}$. Since $\mathfrak{M} \subset \mathfrak{N}$ is given, it suffices to show that $\mathfrak{N} \subset \mathfrak{M}$. Thus, we prove that if $[n,0] \in \mathfrak{N}$, then $[n,0] \in \mathfrak{M}$ for each $n \in N$. To this end, set

$$M = \{m \mid m \in N \text{ and } [m,0] \in \mathfrak{M}\}.$$

Assume, for the moment, we have proved $M = N$. Then for each $x \in N$, $[x,0] \in \mathfrak{M}$. Now let $[z,0] \in \mathfrak{N}$; then $z \in N = M$, hence $[z,0] \in \mathfrak{M}$. Consequently, $\mathfrak{N} \subset \mathfrak{M}$, and as remarked above, this yields $\mathfrak{N} = \mathfrak{M}$, the desired result.

Thus we have to prove that $M = N$. By definition, $M \subset N$; therefore we need only prove $0 \in M$ and M is hereditary. Clearly $0 \in M$;

for $[0,0] \in \mathfrak{M}$ by hypothesis.   Suppose $n \in M$, i.e., suppose $[n,0] \in \mathfrak{M}$. By hypothesis, $[n + 1, 0] \in \mathfrak{M}$, and since $n + 1 \in N$, it follows that $n + 1 \in M$.   Therefore M is hereditary and the theorem is proved.

q.e.d.∗

EXERCISES

1. In proving Theorem 11, the PFI for the natural numbers was used. Give an alternate proof for Theorem 11 using the WOP for the natural numbers in place of the PFI.

2. Let S be a set of integers containing a smallest integer, $[s,0]$; thus $[s,0] \leq' [x,0]$ for all $[x,0] \in S$.   Prove: if $[a + 1, 0] \in S$ whenever $[a,0] \in S$, then S is the set of all integers $\geq' [s,0]$.   (Note: this statement of the PFI for the integers is what enables us to "start" an induction with $n = 1$.   This is usually the beginning of an induction in elementary algebra.)

3. Prove the following generalization of Exercise 2:

Let S be a nonempty set of integers containing a smallest integer. If $[a + 1, b] \in S$ whenever $[a,b] \in S$, then S is the set of all integers greater than or equal to the smallest integer in S.   (How does this theorem compare with Exercise 2?)

∗ Theorem 12. The Well-Ordering Principle holds for the nonnegative integers; namely, every nonempty set of nonnegative integers contains a smallest integer.

proof: To prove this theorem, we shall rely upon the WOP for the natural numbers and we use, again, the one-one correspondence $f(x) = [x,0]$, $x \in N$.

Let $\mathfrak{M} \neq \phi$ be a subset of $\mathfrak{N}$; we wish to prove that $\mathfrak{M}$ contains a smallest element.   Set $M = \{m \mid f(m) = [m,0] \in \mathfrak{M}\}$.   Since $\mathfrak{M} \neq \phi$ and f is one-one, then $M \neq \phi$.   Further, by definition, $M \subset N$. Since M is a nonempty subset of the natural numbers by the WOP for N, it follows that M contains a smallest element, say s.   Then $f(s) = [s,0] \in \mathfrak{M}$; we contend that $[s,0]$ is a smallest element of $\mathfrak{M}$.

Let $[y,0]$ be any element in $\mathfrak{M}$.   Then $y \in M$ and, since s is the smallest element of M, $s \leq y$.   Hence (Definition 9 and Theorem 10), $f(s) \leq' f(y)$, i.e., $[s,0] \leq' [y,0]$.

q.e.d.∗

EXERCISES

1. Prove Theorem 12 from Theorem 11 without using the WOP for the natural numbers.

2. Prove Theorem 11 from Theorem 12 without using the PFI for the natural numbers.

We are now in a position to simplify the notations for the integers and bring them into the familiar forms. For the purposes of this discussion (and the exercises immediately following) let m, n be variables whose range is the set N of natural numbers, and let x, y be variables whose range is the set Z of integers.

Since every nonnegative integer has a unique representation,

(3.7) $$[n,0],$$

where $n \in N$, we shall drop the symbol (3.7) completely, henceforth denoting the integer by the symbol,

$$n.$$

However, in using this abbreviation, it must be clear from the context (or else should be stated explicitly) whether "n" refers to a natural number or whether it is an abbreviation for the integer denoted by "$[n,0]$." Thus, if one speaks of "the number, 5" it should be clear, either from the context or from an explicit statement, whether a natural number or an integer is under discussion.

Since the inverse of an integer $[n,0]$ is denoted by "$-[n,0]$" and since $[n,0]$ is henceforth to be denoted by "n," consistency demands that the inverse of $[n,0]$ be denoted by "$-n$." On the other hand, we already know that the inverse of $[n,0]$ is $[0,n]$. Hence it follows that

$$-n = [0,n], n \in N.$$

For example, "-5" denotes the negative integer $[0,5] = -[5,0]$.

Finally, we replace the inequality symbols "$<'$" and "$>'$" by the customary ones "$<$" and "$>$" respectively. Thus, if x, y are integers and if $x <' y$, we shall write "$x < y$."

The list of changes of notation is:

> "n" is to replace "$[n,0]$";
> "$-n$" is to replace "$[0,n]$";
> "$x < y$" is to replace "$x <' y$";
> "$x > y$" is to replace "$x >' y$";
> "$x + y$" is to replace "$x \oplus y$";
> "xy" is to replace "$x \odot y$".

EXERCISES

Prove: for all integers x, y:

1. If x is a positive integer, $-x$ is a negative integer; if x is negative, then $-x$ is positive.

2. $x(-y) = (-x)y = -(xy)$.

3. $(-x)(-y) = xy$.   Hence, $(-1)(-1) = 1$.

4. $(-1)x = -x$, $-(-x) = x$.

5. $-(x + y) = (-x) + (-y)$.

6. If $x \cdot y = 0$, then $x = 0$ or $y = 0$.

7. Restate Theorems 11 and 12 in terms of the new notation.

8. Prove that if a is an integer, then there is no integer x such that $a < x < a + 1$.   In particular, there is no integer y such that $0 < y < 1$.

3.3.   NUMBER-THEORETIC PROPERTIES OF THE INTEGERS: GENERALIZED OPERATIONS

We turn to a study of a few elementary number-theoretic properties of the integers, some of which may already be familiar to the reader. An indispensable tool for this study is the concept of "generalized operation" already introduced in connection with the natural numbers. Theorem 62 (Chapter 2) is sufficiently powerful to yield the desired results.

* **Definition 10.**   For each $y \in Z$ the function $m_y : Z \longrightarrow Z$ defined by

$$m_y(x) = x \cdot y, \ x \in Z$$

is the *multiplication function for y;* the function $s_y : Z \longrightarrow Z$ defined by

$$s_y(x) = x + y, \ x \in Z$$

is the *addition function for y.**

These are not the same functions as were defined in Chapter 2, Definitions 27 and 29, respectively.

* Now, in Theorem 62 (Chapter 2) let us take $A = Z$, $\mathcal{K} =$ the set of multiplication functions, $\varphi: N_1 \longrightarrow \mathfrak{K}$ a sequence of multiplication functions, $a = 1$. Then Theorem 62 (Chapter 2) yields as an immediate

corollary 1: Let φ be a sequence of multiplication functions. Then there exists one and only one function $\pi_\varphi: N \longrightarrow Z$ such that

(3.8)
$$\begin{cases} \pi_\varphi(0) = 1, \\ \pi_\varphi(n + 1) = m_{\varphi(n+1)}(\pi_\varphi(n)), \end{cases}$$

for all $n \in N$. π_φ is the *generalized product function* for the sequence φ. It is said to be *defined recursively* by equations (3.8). As in Chapter 2, the symbol "$\prod\limits_{i=1}^{n} \varphi(i)$" is defined by

$$\prod_{i=1}^{n} \varphi(i) = \pi_\varphi(n).$$

Similarly, if in Theorem 62, Chapter 2, we take $A = Z$, $\mathfrak{K} =$ the set of addition functions, $\varphi: N_1 \longrightarrow \mathfrak{K}$ a sequence of addition functions and $a = 0$, we obtain

corollary 2: For each sequence φ of addition functions there exists one and only one function $\sigma_\varphi: N \longrightarrow Z$ such that

(3.9)
$$\begin{cases} \sigma_\varphi(0) = 0, \\ \sigma_\varphi(n + 1) = s_{\varphi(n+1)}(\sigma_\varphi(n)), \end{cases}$$

for all $n \in N$. σ_φ is the *generalized sum function* for the sequence φ. It is said to be *defined recursively* by equations (3.9). The symbol "$\sum\limits_{i=1}^{n} \varphi(i)$" is defined by

$$\sum_{i=1}^{n} \varphi(i) = \sigma_\varphi(n).$$

Finally, as in the case of the natural numbers, one can prove generalized associative and commutative laws with respect to $+$ and $\cdot$ for Z. These results we accept without further discussion.∗

Before we begin the study of some elementary number-theoretic properties, it is convenient to have the concept of the *absolute value function*.

∗ **Definition 11.** The *absolute value function on Z* is the function

$$|\,| = \{(x,x) \mid x \in Z \text{ and } x \geq 0\} \cup \{(x,-x) \mid x \in Z \text{ and } x < 0\}.$$

In the customary functional notation, this definition can be stated as

$$|x| = \begin{cases} x, & \text{if } x \in Z \text{ and } x \geq 0, \\ -x, & \text{if } x \in Z \text{ and } x < 0. \end{cases}$$

Clearly, $\mathfrak{D}(||) = Z$, $\mathfrak{R}(||) \subset Z$ and for all x, $y \in Z$ if $x = y$ then $|x| = |y|$. Therefore, $||$ is indeed a function on Z.∗

The required properties of $||$ are few in number and sufficiently simple so that the proofs of most can be left as exercises. We verify only 6, 7 and 8 below:

*1. For each $x \in Z$, $|x| = |-x|$; $x \leq |x|$ and $-|x| \leq x$.
2. If $|x| = |y|$, where $x, y \in Z$, then $x = \pm y$.
3. For all $x, y \in Z$, $|xy| = |x| \cdot |y|$.
4. For each $x \in Z$, if $a \in Z$, $a \geq 0$, then $|x| \leq a$ if and only if $-a \leq x \leq a$.
5. For all $x, y \in Z$, $|x| \leq |y|$ if and only if $-|y| \leq x \leq |y|$.
6. For all $x, y \in Z$, $|x + y| \leq |x| + |y|$. (This is the *triangle inequality*.)

proof: By 1, $-|x| \leq x \leq |x|$ and $-|y| \leq y \leq |y|$, so that

$$-(|x| + |y|) \leq x + y \leq |x| + |y|.$$

Hence, by 4, taking $|x| + |y|$ in place of a and $x + y$ in place of x, it follows that

$$|x + y| \leq |x| + |y|.$$

7. For all $x, y \in Z$, $|x - y| \leq |x| + |y|$.

proof: This is obtained from 6 by replacing y by $-y$ and using the fact that $|y| = |-y|$.

8. For all $x_1, x_2, \ldots, x_n \in Z$, where $n \in N$, $|x_1 + x_2 + \ldots + x_n| \leq |x_1| + |x_2| + \ldots + |x_n|$.

proof: Use 6 and induction.∗

Now, to the number-theoretic properties: these are not only useful in the sequel, but also have considerable intrinsic interest.

* Definition 12.   Let $a, b \in Z$, $b \neq 0$. If there is an integer c such that $a = bc$, then b *divides* a (b is a *divisor* of a; b is a *factor* of a; a is a *multiple* of b); we write "$b \mid a$." If b does not divide a, i.e., if there is no integer c such that $a = bc$, we write "$b \nmid a$."∗

EXERCISES

Prove:

1. If $a \neq 0$ and $b \neq 0$ and if $b \mid a$ then $|b| \leq |a|$. (Hence, if $a > 0$ and $b > 0$ and $b \mid a$ then $b \leq a$.)

2. $b \mid a$ if and only if $-b \mid a$. Hence, $b \mid a$ if and only if $|b| \mid |a|$.

3. The only divisors of 1 are ± 1.

4. If $a \mid b$ and $b \mid a$ then $a = \pm b$.

An important theorem of elementary number theory is the famous division algorithm of Euclid:

* Theorem 13. If $a, b \in Z$, $b \neq 0$, then there are unique integers q and r such that

(3.10) $$a = qb + r$$

where $0 \leq r < |b|$. (q is the *quotient*; r is the *remainder*.)

proof: We begin by showing that the set

$$S = \{a - x \cdot |b| \mid x \in Z\}$$

contains a nonnegative integer. For, if $a \geq 0$, then

$$a = a - 0 \cdot |b| \in S.$$

On the other hand, suppose $a < 0$. Then $-a > 0$; since $|b| > 0$, $|b| \geq 1$. Hence, $(-a) \cdot |b| \geq -a$, and therefore $a - a \cdot |b| \geq 0$. Clearly, $a - a \cdot |b| \in S$. Thus, in every case, S contains a nonnegative integer.

Next, let T be the subset of S defined by

$$T = \{z \mid z \in S \text{ and } z \geq 0\}.$$

By the foregoing, $T \neq \phi$, and so by the WOP, T contains a smallest integer r. Consequently, there is an integer y such that

$$r = a - y \cdot |b|, 0 \leq r.$$

We claim that $r < |b|$. Otherwise, $r \geq |b|$ and then

$$a - (y + 1) \cdot |b| = (a - y \cdot |b|) - |b| = r - |b| \in T.$$

But $r - |b| < r$; and since r is the smallest integer in T, this is a contradiction.   Therefore, $r < |b|$.

To establish the uniqueness of y and r suppose that

$$a = y_1 \cdot |b| + r_1 = y_2 \cdot |b| + r_2$$

where $0 \le r_1 < |b|$, $0 \le r_2 < |b|$.   We may assume that $r_1 \le r_2$. Then $0 \le r_2 - r_1 < |b|$ and we have

$$r_2 - r_1 = (y_1 - y_2) \cdot |b|,$$

hence $|b|$ is a divisor of the nonnegative integer $r_2 - r_1$.   Now, if $r_2 - r_1 > 0$, then by Exercise 1, page 150, $|b| \le r_2 - r_1$, a contradiction. Therefore $r_2 - r_1 = 0$, $r_1 = r_2$ and $y_1 = y_2$.

Finally, if $b > 0$ then $|b| = b$ and we take $q = y$.   If $b < 0$ then $|b| = -b$ and we take $q = -y$.   In either case the integers q and r are such that (3.10) holds.

q.e.d.

Definition 13(a).   Let a, $b \in Z$.   An integer d is a *common divisor* of a and b if and only if $d \mid a$ and $d \mid b$.

Definition 13(b).   Let a, $b \in Z$, not both zero.   An integer d is a *greatest common divisor* of a and b if and only if

   (i) d is a common divisor of a and b; and
   (ii) if $c \mid a$ and $c \mid b$ then $c \mid d$.

Suppose d is a greatest common divisor of a and b.   By Exercise 2, page 150, it follows that $-d$ is also a greatest common divisor of a and b.

Definition 13(c).   Let d be a greatest common divisor of a and b.   Then $|d|$ is *the* greatest common divisor of a and b.   It is denoted by

$$\text{g.c.d. } (a,b).$$

Definition 14.   The integers a, b are *relatively prime* (or, *coprime*) if and only if g.c.d. $(a,b) = 1$.

We now prove that every pair of integers, not both zero, has a g.c.d. The first step is a

lemma: Let T be a nonempty subset of Z such that

$$\text{for all } x, y \in T, x + y \in T \text{ and } x - y \in T.$$

Then T contains a smallest nonnegative integer d such that

(3.11) $$T = \{nd \mid n \in Z\}.$$

If $T \neq \{0\}$, then $d > 0$.

proof: If $T = \{0\}$ then $d = 0$ does the trick.

Suppose $T \neq \{0\}$; we show, first, that T contains a smallest positive integer.

T contains a positive integer. For, since $T \neq \{0\}$, there is an integer $x \neq 0$ in T. By hypothesis, $0 = x - x \in T$. Hence, $-x = 0 - x \in T$. Since one of the two integers x, $-x$ is positive, it follows that T contains a positive integer.

Next, set

$$T^+ = \{u \mid u \in T \text{ and } u > 0\};$$

by the foregoing $T^+ \neq \phi$. By the WOP for the integers, T^+ contains a smallest integer $d > 0$; in fact, d is the smallest positive integer in T.

We now leave it to the reader to prove:

(a) $\{nd \mid n \in Z \text{ and } n \geq 0\} \subset T$ (hint: use the PFI);
(b) hence, $\{nd \mid n \in Z\} \subset T$.

Assuming (b) we now show that every element in T is an integral multiple of d. This result, together with (b), yields (3.11).

Suppose there is a $y \in T$ which is not an integral multiple of d. By Theorem 13 there are integers q and r such that

$$y = qd + r, \ 0 \leq r < |d| = d.$$

Now $r > 0$; otherwise $y = qd$, a contradiction. But then

$$r = y - qd,$$

and since $y \in T$, $qd \in T$ (by (b)), $r = y - qd \in T$. So $0 < r < d$ and $r \in T$; this is a contradiction since d is the smallest positive integer in T. Hence, every element in T is an integral multiple of d, and the lemma is proved.

Theorem 14. If a, b are integers not both zero, then the g.c.d. (a,b) exists. In fact, there exist integers s, t such that

$$\text{g.c.d. } (a,b) = sa + tb.$$

proof: Set $T = \{xa + yb \mid x, y \in Z\}$; clearly $T \neq \phi$, and if $u, v \in T$ then $u + v$ and $u - v$ are in T. Hence, by the lemma, T contains a smallest integer $d > 0$ such that $T = \{nd \mid n \in Z\}$. And by definition of T there exist integers s and t such that

$$d = sa + tb.$$

We claim that $d = $ g.c.d. (a,b).

Since $a = 1 \cdot a + 0 \cdot b \in T$ and $b = 0 \cdot a + 1 \cdot b \in T$, it follows that $d \mid a$ and $d \mid b$ (lemma) so that d is a common divisor of a and b. Suppose $c \mid a$ and $c \mid b$; we claim that $c \mid d$ and therefore (since $d > 0$) $d = $ g.c.d. (a,b). From $c \mid a$ and $c \mid b$ we have $a = cu$, $b = cv$ with $u, v \in Z$, whence

$$\begin{aligned} d &= sa + tb \\ &= s(cu) + t(cv) \\ &= (su + tv)c; \end{aligned}$$

since $su + tv \in Z$, then $c \mid d$.

q.e.d._*

The explicit computation of g.c.d.'s is carried out by repeated applications of the division algorithm. The details are given in the exercises below. Note, first, that if $a = 0$ and $b \neq 0$ then g.c.d. $(a,b) = |b|$. Similarly, if $a \neq 0$, $b = 0$, g.c.d. $(a,b) = |a|$. Therefore we may assume in the exercises below, that $a \neq 0 \neq b$.

EXERCISES

1. Prove: For all $a, b \in Z$, g.c.d. $(a,b) = $ g.c.d. $(-a,b) = $ g.c.d. $(a,-b) = $ g.c.d. $(-a,-b)$.

2. Assume a, b both positive and $a \nmid b$ and $b \nmid a$. Define $q_i, r_i \in Z$, $i \in N, i \geq 1$ by:

$$\begin{aligned} a &= q_1 b + r_1, \ 0 \leq r_1 < b, \\ b &= q_2 r_1 + r_2, \ 0 \leq r_2 < r_1, \\ r_i &= q_{i+2} r_{i+1} + r_{i+2}, \ i \in N, i \geq 1. \end{aligned}$$

Prove: There exists a natural number n such that $r_{n+1} = 0$.

3. By Exercise 2 there is a smallest natural number m such that $r_{m+1} = 0$. Prove that g.c.d. $(a,b) = $ g.c.d. $(b,r_1) = $ g.c.d. (r_i,r_{i+1}), $1 \leq i \leq m - 1$, hence that $r_m = $ g.c.d. (a,b).

4. Use the method of the above exercises to compute g.c.d. $(991,236)$.

* Definition 15. An integer $p \neq 0, \pm 1$ is *prime* if and only if $d \mid p$ implies $d = \pm 1$ or $d = \pm p$.*

EXERCISE

Prove that 2, 3 and 5 are primes.

* Theorem 15. If p is prime and $p \mid ab$ then $p \mid a$ or $p \mid b$.

proof: We assume that $p \nmid a$ and deduce $p \mid b$. (Note that this is the kind of reasoning discussed in connection with Theorem 10, Chapter 2). Since $p \nmid a$ and p is prime, g.c.d. $(p,a) = 1$. By Theorem 14 there exist integers s and t such that

$$1 = sp + ta.$$

Then

$$b = spb + tab.$$

Since $p \mid ab$ there is an integer $c \in Z$ such that $ab = cp$. Hence,

$$b = spb + tcp = (sb + tc)p,$$

where $sb + tc \in Z$. Therefore $p \mid b$.

q.e.d.

corollary 1: If p is prime and $p \mid \prod_{i=1}^{n} a_i$, $n \geq 1$, then $p \mid a_i$ for some i such that $1 \leq i \leq n$.

proof: Exercise.

corollary 2: If $c \mid ab$ and a, c are relatively prime, then $c \mid b$

proof: Exercise.

Theorem 16. (Unique Factorization Theorem)

(α) If a is an integer, $a \neq 0, \pm 1$, then there exists a natural number $r \geq 1$ and positive primes $p_1, p_2, \ldots, p_r$ such that

$$a = \pm 1 \cdot p_1 \cdot p_2 \cdot \ldots \cdot p_r.$$

(β) If $p_1 \cdot p_2 \cdot \ldots \cdot p_r = q_1 \cdot q_2 \cdot \ldots \cdot q_s$ where the p's and q's are positive primes, then $r = s$ and each $p_i = $ some q_j.

(α) and (β) are summed up by:

Every integer $a \neq 0$, ± 1 is $+1$ or -1 times a product of positive primes which is unique to within order of the primes and factors $+1$ or -1.

Theorem 16 is also called the "Fundamental Theorem of Arithmetic."

proof: We begin by proving the theorem for the positive integers; the extension to the negative integers will be exceedingly simple.

(α) Let $T = \{a \mid a \in Z, a > 1$ and a is not a product of positive primes$\}$. In other words, $a \in T$ if and only if

(i) a is an integer, $a > 1$, and

(ii) for all sets $\{p_1, p_2, \ldots, p_r \mid r \in N, r \geq 1$ and $p_1, p_2, \ldots, p_r$ are positive primes$\}$, $a \neq p_1 \cdot p_2 \cdot \ldots \cdot p_r$.

The first part of the theorem will be proved (for the positive integers) if we can show that $T = \phi$.

Suppose $T \neq \phi$. By the WOP for the integers, T contains a smallest integer $m > 0$. Since $m \in T$, there exist integers $a > 1$, $b > 1$ such that $m = a \cdot b$. Then $a < m$ and $b < m$ so that $a \notin T$ and $b \notin T$. Consequently,

$$a = p_1 \cdot p_2 \cdot \ldots \cdot p_r, \quad b = q_1 \cdot q_2 \cdot \ldots \cdot q_s$$

where r, s are natural numbers ≥ 1 and the p's and q's are positive primes. Therefore,

$$m = p_1 \cdot p_2 \cdot \ldots \cdot p_r \cdot q_1 \cdot q_2 \cdot \ldots \cdot q_s,$$

a contradiction. Hence, $T = \phi$ and (α) is proved.

(β) Let

$$S = \{a \mid a \in Z, a > 1, a = p_1 \cdot p_2 \cdot \ldots \cdot p_r = q_1 \cdot q_2 \cdot \ldots \cdot q_s$$

where the p's and q's are positive primes and $r \neq s$ or some $p_i \neq q_j$, for all $j = 1, 2, \ldots, s\}$.

Our object is to prove that $S = \phi$.

If $S \neq \phi$, S contains a smallest integer $u > 0$ (WOP). Thus

(3.12) $$u = p_1 \cdot p_2 \cdot \ldots \cdot p_m = q_1 \cdot q_2 \cdot \ldots \cdot q_n,$$

where m, n are natural numbers ≥ 1, the p's and q's are positive primes, and $m \neq n$ or some $p_i \neq q_j$, for all $j = 1, 2, \ldots, n$. Now

$p_1 \mid p_1 \cdot p_2 \cdot \ldots \cdot p_m$, therefore $p_1 \mid q_1 \cdot q_2 \cdot \ldots \cdot q_n$. By Corollary 1 to Theorem 15, $p_1 \mid$ some q_j. By renumbering the q's, we may assume that $p_1 \mid q_1$. Since p_1 and q_1 are both positive primes, it follows that $p_1 = q_1$. By Theorem 5, part (iii), (cancellation law for the multiplication of integers) we see that

$$v = p_2 \cdot p_3 \cdot \ldots \cdot p_m = q_2 \cdot q_3 \cdot \ldots \cdot q_n,$$

and $0 < v < u$. Therefore, $v \notin S$ (remember, u is the smallest integer in S). Consequently, $m = n$ and each $p_i =$ some q_j, for $i, j \geq 2$; since $p_1 = q_1$, the same is true of the factorizations (3.12). Therefore we have a contradiction and S must be empty.

This completes the proof of the theorem for the positive integers.

Now suppose a is a negative integer, $a \neq -1$. Then $-a$ is positive, $-a > 1$, and by the foregoing there is a unique set $\{p_1, p_2, \ldots, p_r\}$ of positive primes such that

$$-a = p_1 \cdot p_2 \cdot \ldots \cdot p_r.$$

Hence,

$$a = -p_1 \cdot p_2 \cdot \ldots \cdot p_r.$$

Thus, in every case, if $a \neq 0, \pm 1$, there is a unique set $\{p_1, p_2, \ldots, p_r\}$ of positive primes such that

$$a = \pm 1 \cdot p_1 \cdot p_2 \cdot \ldots \cdot p_r.$$

q.e.d.*

(See Exercise 9, page 157.)

EXERCISES

1. Let a, b be nonzero integers and let $d = $ g.c.d. (a,b); then there exist integers x, y such that $a = xd$ and $b = yd$. Prove that g.c.d. $(x,y) = 1$.

2. If $a \neq 0$, prove that g.c.d. $(ab,ac) = |a| \cdot$ g.c.d. (b,c).

3. If a, b, c are nonzero integers, define

$$\text{g.c.d. } (a,b,c) = \text{g.c.d. } (\text{g.c.d. } (a,b),c).$$

Prove: g.c.d. $(a,b,c) = $ g.c.d. $(\text{g.c.d. } (a,c),b) = $ g.c.d. $(\text{g.c.d. } (b,c),a)$.

4. If a, b, c are nonzero integers, prove that there exist integers r, s, t such that g.c.d. $(a,b,c) = ra + sb + tc$.

5. Generalize the definition of g.c.d. for the case of n integers, $n \geq 2$. Prove that if $a_1, a_2, \ldots, a_n, n \geq 2$ are nonzero integers, then there exist integers $r_1, r_2, \ldots, r_n$ such that g.c.d. $(a_1, a_2, \ldots, a_n) = r_1 a_1 + r_2 a_2 + \ldots + r_n a_n$.

6. Let a, b be nonzero integers. An integer m is a *least common multiple* of a and b if and only if it satisfies the following conditions:

1. $a \mid m$ and $b \mid m$, and if
2. n is an integer such that $a \mid n$ and $b \mid n$, then $m \mid n$.

Clearly, if m is a least common multiple of a and b, so is $-m$. Hence, a and b have a positive least common multiple which we call *the* least common multiple of a and b, and denote by "l.c.m. (a,b)." Prove: $|ab| = $ (g.c.d. (a,b)) $\cdot$ (l.c.m. (a,b)).

7. Prove that there exist infinitely many primes.

8. Prove part (α) of Theorem 16 by means of the second PFI.

9. Let $p_{i_1} \cdot p_{i_2} \cdot \ldots \cdot p_{i_r}$ be a product of primes. If $p_{i_1} = p_{i_2} = \ldots = p_{i_k} = p$ and if, for all $j \neq i_1, i_2, \ldots, i_k$, where $j = 1, 2, \ldots, r$, $p_j \neq p$, then the prime p occurs in $p_1 \cdot p_2 \cdot \ldots \cdot p_r$ with *multiplicity* k. Prove:

If $p_1 \cdot p_2 \cdot \ldots \cdot p_r = q_1 \cdot q_2 \cdot \ldots \cdot q_s$ are products of positive primes then each p_i occurs in the two products with the same multiplicity.

3.4.   THE RATIONAL NUMBERS

We turn now to the construction of a system of numbers, the rationals, in which unrestricted division, except by zero, may be carried out. The procedure we use here parallels very closely that used for the construction of the integers. Hence, our definitions and theorems will fit into a pattern similar to that of Section 3.2. In particular we shall find that the rationals contain a copy of the integers. In the following, we shall denote integers by lower case Latin letters.

* Definition 16. $Z_0 = Z - \{0\}$.

Definition 17. If (a,b), $(c,d) \in Z \times Z_0$, then $(a,b) \sim (c,d)$ if and only if $ad = cb$. An element $(a,b) \in Z \times Z_0$ is a *fraction*. The left component a is the *numerator*, the right component b is the *denominator* of the fraction.

Alternatively, the definition of $\sim$ can be stated:

$\sim$ is the subset of $(Z \times Z_0) \times (Z \times Z_0)$ consisting of all elements $((a,b), (c,d))$ such that
$$ad = cb.$$

Theorem 17. $\sim$ is an equivalence relation on $Z \times Z_0$.

proof: Clearly, $\sim$ is reflexive and symmetric. To establish transitivity, suppose $(a,b) \sim (c,d)$ and $(c,d) \sim (e,f)$. Then $ad = cb$ and $cf = ed$; we wish to prove that $af = eb$. From $ad = cb$ we deduce $adf = cbf$; and from $cf = ed$ we have $bcf = bed$, or $cbf = ebd$. Hence,

$$adf = afd = ebd$$

and by the cancellation law for the multiplication of integers (remember, d is not 0),

$$af = eb.$$

Therefore, by Definition 17,

$$(a,b) \sim (e,f).$$

q.e.d.

By Theorem 12, Chapter 1, $\sim$ determines a partition R of $Z \times Z_0$. The elements of R are denoted by

$$\frac{a}{b},$$

where

$$\frac{a}{b} = \{(x, y) \mid (x, y) \in Z \times Z_0 \text{ and } (x, y) \sim (a, b)\}.$$

The elements $a/b \in R$ are the equivalence classes with respect to the equivalence relation $\sim$.

Definition 18. The elements $a/b \in R$ are *rational numbers*. R is the set of rational numbers. The symbol "a/b" is read "*a* over *b*."*

EXAMPLES

1. $\frac{2}{3} = \{(x,y) \mid (x,y) \in Z \times Z_0 \text{ and } (x,y) \sim (2,3)\}$. Thus, among the elements in the equivalence class $\frac{2}{3}$ are the fractions $(2,3)$, $(4,6)$, $(6,9)$, $(-2,-3)$, $(-10,-15)$, etc.

2. $\frac{5}{1} = \{(x,y) \mid (x,y) \in Z \times Z_0$ and $(x,y) \sim (5,1)\}$. Therefore $(5,1) \in \frac{5}{1}$, $(10,2) \in \frac{5}{1}$, $(-54,-27) \in \frac{5}{1}$, etc.

Before introducing addition and multiplication for rational numbers, we need results analogous to Theorems 2 and 3 for the integers.

*Theorem 18. $a/b = c/d$ if and only if $ad = cb$.

proof: Definition of rational number.

Theorem 19.   If $a/b = c/d$ and $e/f = g/h$ then

(i) $\dfrac{af + eb}{bf} = \dfrac{ch + gd}{dh}$, and

(ii) $ae/bf = cg/dh$.

proof: To prove (i), we show that

$$(af + eb)dh = (ch + gd)bf.$$

We have

$$(af + eb)dh = (af)(dh) + (eb)(dh)$$
$$= (ad)(fh) + (eh)(bd).$$

By hypothesis, $a/b = c/d$ and $e/f = g/h$ so that $ad = cb$ and $eh = gf$. Hence,

$$(ad)(fh) + (eh)(bd) = (cb)(fh) + (gf)(bd) = (ch + gd)bf$$

and therefore

$$(af + eb)dh = (ch + gd)bf.$$

We leave part (ii) of the proof as an exercise.

$$\text{q.e.d.}$$

corollary: The sets

(3.13) $$\perp = \left\{ \left(\left(\frac{a}{b}, \frac{e}{f} \right), \frac{af + eb}{bf} \right) \middle| \frac{a}{b}, \frac{e}{f} \in R \right\}$$

and

(3.14) $$\circledast = \left\{ \left(\left(\frac{a}{b}, \frac{e}{f} \right), \frac{ae}{bf} \right) \middle| \frac{a}{b}, \frac{e}{f} \in R \right\}$$

are binary operations on R. In other words, both $\perp$ and $\circledast$ are mappings of $R \times R$ into R.

proof: The proof parallels closely that of the Corollary to Theorem 3.

Consider first the set $\perp$; clearly $\mathfrak{D}(\perp) = R \times R$ and $\mathfrak{R}(\perp) \subset R$. Therefore we need only show that if $a/b = c/d$ and $e/f = g/h$ then

$$\frac{af + eb}{bf} = \frac{ch + gd}{dh}.$$

But this result is assured by part (i) of Theorem 19. Hence, $\perp$ is a binary operation on R.

In a similar fashion, $\mathfrak{D}(\circledast) = R \times R$, $\mathfrak{R}(\circledast) \subset R$ and by (ii) of Theorem 19, $\circledast$ is a mapping of $R \times R$ into R. Consequently, $\circledast$ is a binary operation on R.

q.e.d.

Definition 19. The binary operations $\perp$ and $\circledast$ are *sum* and *product*, respectively, of rational numbers.

To bring the notation closer to customary usage, we write:

(3.15) $$\frac{a}{b} \perp \frac{e}{f} = \frac{af + be}{bf}, \quad \frac{a}{b}, \frac{e}{f} \in R,$$

in place of (3.13), and

(3.16) $$\frac{a}{b} \circledast \frac{e}{f} = \frac{ae}{bf}, \quad \frac{a}{b}, \frac{e}{f} \in R,$$

instead of (3.14). The "$+$" and "$\cdot$" on the right sides of (3.15) and (3.16) denote sum and product, respectively, of *integers;* the "$\perp$" and "$\circledast$" on the left sides of these equations denote sum and product, respectively, of rational numbers.*

The following examples illustrate that $\perp$ and $\circledast$ are in agreement with the elementary concepts of sum and product, respectively, of rational numbers.

EXAMPLES

1. $\dfrac{1}{2} \perp \dfrac{2}{3} = \dfrac{1 \cdot 3 + 2 \cdot 2}{2 \cdot 3} = \dfrac{7}{6};$

2. $\dfrac{-2}{5} \perp \dfrac{3}{7} = \dfrac{(-2) \cdot 7 + 5 \cdot 3}{5 \cdot 7} = \dfrac{1}{35};$

3. $\dfrac{2}{-5} \perp \dfrac{3}{7} = \dfrac{2 \cdot 7 + (-5) \cdot 3}{(-5) \cdot 7} = \dfrac{-1}{-35} = \dfrac{1}{35}$;

4. $\dfrac{5}{8} \circledast \dfrac{2}{3} = \dfrac{5 \cdot 2}{8 \cdot 3} = \dfrac{10}{24} = \dfrac{5}{12}$;

5. $\dfrac{-5}{8} \circledast \dfrac{2}{3} = \dfrac{(-5) \cdot 2}{8 \cdot 3} = \dfrac{-10}{24} = \dfrac{-5}{12}$;

6. $\dfrac{5}{-8} \circledast \dfrac{2}{3} = \dfrac{5 \cdot 2}{(-8) \cdot 3} = \dfrac{10}{-24} = \dfrac{5}{-12}$;

7. $\dfrac{-5}{8} \circledast \dfrac{-2}{3} = \dfrac{(-5) \cdot (-2)}{8 \cdot 3} = \dfrac{10}{24} = \dfrac{5}{12}$;

8. $\dfrac{-5}{8} \circledast \dfrac{2}{-3} = \dfrac{(-5) \cdot 2}{8 \cdot (-3)} = \dfrac{-10}{-24} = \dfrac{5}{12}$.

The reader should construct more examples to assure himself of the complete agreement between the theory presented here and the practice of elementary arithmetic.

3.5.   THE ARITHMETIC OF THE RATIONAL NUMBERS

* Theorem 20.   (i) $a/b = c/b$ if and only if $a = c$; if $a \neq 0$ then $a/b = a/d$ if and only if $b = d$.

(ii) For each rational number a/b there is a unique smallest positive integer n and a unique integer m such that $a/b = m/n$; and

(iii) if $m \neq 0$ then m and n are relatively prime;

(iv) further, there is a nonzero integer x such that $a = mx$ and $b = nx$; and finally

(v) $a/b = \{(my, ny) \mid y \in Z_0\}$.

proof: (i) By Theorem 18, $a/b = c/b$ if and only if $ab = cb$.   By the cancellation law for the multiplication of integers, since $b \neq 0$, it follows that $a = c$.   Conversely, if $a = c$, it is clear that $a/b = c/b$.   The second part of (i) is left as an exercise.

(ii) Let

$$S = \{y \mid y \in \mathfrak{N} \text{ and there is an } x \in Z \text{ such that } (x, y) \in a/b\}.$$

Since $S \subset \mathfrak{N}$ and since $S \neq \phi$ (why?) it follows, by the WOP, that S contains a smallest, positive integer n.   Thus, n is the denominator of a fraction $(m, n) \in a/b$.   If also $(m_1, n) \in a/b$ then $(m_1, n) \sim (m, n)$,

$m_1n = mn$ and since $n \neq 0$, $m_1 = m$. Finally, since $(m,n) \in a/b$, $(m,n) \sim (a,b)$; therefore $mb = an$ whence, by Theorem 18, $m/n = a/b$.

(iii) To prove m, n relatively prime (where $m \neq 0$), suppose that $q \mid m$ and $q \mid n$; we may assume that q is positive (why?). We prove that $q = 1$. For, suppose $q \neq 1$. Then there exist integers x and y such that $m = qx$ and $n = qy$. Since n and q are positive, so is y, and by Exercise 1 of page 150, $y \leq n$. Moreover, since $q \neq 1$, we have $y < n$. But $(m,n) = (qx,qy) \sim (x,y)$ so that $(x,y) \in m/n$; this is a contradiction since (m,n) is the fraction with the smallest positive denominator in its equivalence class.

(iv) With a, b, m, n as in (ii), we show there is a nonzero integer x such that $a = mx$ and $b = nx$. If $a = 0$, then $m = 0$ and $n = 1$ (why?). Taking $x = b$, we have

$$a = 0 \cdot b,$$
$$b = 1 \cdot b,$$

as claimed.

Suppose $a \neq 0$. Since $a/b = m/n$ we have $an = mb$, whence $m \mid an$. Since m and n are relatively prime, it follows (Corollary 2, Theorem 15) that $m \mid a$; hence there is an integer $x \neq 0$ such that $a = mx$. On the other hand, since $n \mid mb$ and since n and m are relatively prime, $n \mid b$. Therefore $b = ny$ where y is a nonzero integer. Thus

$$an = nmx, \quad bm = nmy,$$
and since $an = mb$,
$$nmx = nmy.$$

By the cancellation law for the multiplication of integers (since $nm \neq 0$), we deduce

$$x = y.$$

This concludes the proof of (iv) in the case $a \neq 0$.

(v) Exercise.

q.e.d.

Definition 20. Suppose n is the smallest positive integer such that $a/b = m/n$. Then $(a,b) \sim (m,n)$ and (m,n) is *the reduction of the fraction (a,b) to lowest terms*.

Theorem 21. If $a/b = m/n$ and $a \neq 0$, then (m,n) is the reduction of (a,b) to lowest terms if and only if m and n are relatively prime.

proof: Theorem 20.

q.e.d.

Theorem 22.   The rational number $\frac{0}{1}$ is a zero element for R; i.e., for each $a/b \in R$,

$$\frac{a}{b} \perp \frac{0}{1} = \frac{a}{b} \quad \text{and} \quad \frac{a}{b} \circledast \frac{0}{1} = \frac{0}{1}.$$

proof: Theorem 22 follows easily from the definitions of $\perp$ and $\circledast$.

q.e.d.*

EXERCISE

Prove that R has at most one zero element.

* Theorem 23   (i) Sum and product of rationals are commutative and associative;

   (ii) multiplication is distributive over addition;

   (iii) the cancellation laws hold for addition and multiplication.

proof: (i) Exercise.

$$\text{(ii)}\ \frac{a}{b} \circledast \left(\frac{c}{d} \perp \frac{e}{f} \right) = \frac{a}{b} \circledast \frac{cf + ed}{df} = \frac{a(cf + ed)}{bdf}$$

$$= \frac{acf + aed}{bdf}$$

$$= \frac{bacf + baed}{bbdf}$$

$$= \frac{ac}{bd} \perp \frac{ae}{bf}$$

$$= \left(\frac{a}{b} \circledast \frac{c}{d} \right) \perp \left(\frac{a}{b} \circledast \frac{e}{f} \right).$$

(iii) Cancellation law for addition of rationals.   Given $a/b \perp c/d = e/f \perp c/d$ we wish to prove that $a/b = e/f$.   From the hypothesis we have

$$\frac{ad + cb}{bd} = \frac{ed + cf}{fd},$$

or,

$$(ad + cb)fd = (ed + cf)bd,$$
$$adfd + cbfd = edbd + cfbd.$$

By the cancellation law for the addition of integers we have

$$adfd = edbd,$$

and by the cancellation law for the multiplication of integers,

$$af = eb,$$

whence

$$\frac{a}{b} = \frac{e}{f}.$$

The cancellation law for multiplication is left as an exercise.

q.e.d.*

The next theorem shows that subtraction is unrestricted in R.

*Theorem 24. For every pair of rational numbers a/b, c/d there is a unique rational number x/y such that

$$\frac{c}{d} \perp \frac{x}{y} = \frac{a}{b}.$$

In particular, for every rational number c/d there is a unique rational number u/v such that

$$\frac{c}{d} \perp \frac{u}{v} = \frac{0}{1}.$$

proof: We have

$$\frac{c}{d} \perp \frac{ad - bc}{bd} = \frac{(ad - bc)d + (bd)c}{(bd)d}$$

$$= \frac{add}{bdd}$$

$$= \frac{a}{b}.$$

Hence $x/y = (ad - bc)/bd$ is a solution of the first equation. Its uniqueness is a consequence of the cancellation law for the addition of rational numbers. (Details?)

The second statement follows by taking $a = 0, b = 1$. Then

$$u/v = -c/d.$$

q.e.d.

Definition 21(a). For each $c/d \in R$ the unique u/v such that $c/d \perp u/v = 0/1$ is the *additive inverse* or *negative* of c/d. It is denoted by

$$-\frac{c}{d}.$$

corollary 1: $-(c/d) = -c/d = c/-d.$

proof: $-(c/d) = -c/d$ follows from Theorem 24 and Definition 21(a).

Also,

$$\frac{c}{d} \perp \frac{c}{-d} = \frac{c(-d) + cd}{d(-d)} = \frac{0}{d(-d)} = \frac{0}{1}$$

and, again by Theorem 24 and Definition 21(a), $-c/d = c/-d$.

q.e.d.

Definition 21(b). For each a/b and each c/d in R, the unique x/y such that $c/d \perp x/y = a/b$ is *the difference of a/b and c/d.* It is denoted by

$$\frac{a}{b} - \frac{c}{d}.$$

corollary 2:

(i) $\dfrac{a}{b} - \dfrac{c}{d} = \dfrac{ad - bc}{bd} = \dfrac{a}{b} \perp \dfrac{-c}{d} = \dfrac{a}{b} \perp \dfrac{c}{-d}.$

(ii) $\dfrac{0}{1} - \dfrac{c}{d} = -\dfrac{c}{d}.$

proof: Theorem 24, Definition 21 and Corollary 1.

q.e.d.∗

The next step in the development of the arithmetic of R is to show that unrestricted division, except by the zero element, can be carried out in R.

* Theorem 25. 1/1 is a unity element for R; i.e.,

$$\frac{a}{b} \circledast \frac{1}{1} = \frac{a}{b}, \quad \frac{a}{b} \in R.$$

proof: Definition of $\circledast$.

q.e.d.

Theorem 26. For all a/b, c/d in R, $a/b \neq 0/1$, there is a unique $x/y \in R$ such that

$$\frac{a}{b} \circledast \frac{x}{y} = \frac{c}{d}.$$

In particular, there is a unique $u/v \in R$ such that

$$\frac{a}{b} \circledast \frac{u}{v} = \frac{1}{1}.$$

proof: Since $a/b \neq 0/1$, $a \neq 0$, and so b/a is a rational number as is bc/ad. Then

$$\frac{a}{b} \circledast \frac{bc}{ad} = \frac{bca}{bad} = \frac{c}{d}.$$

This shows that $x/y = bc/ad$ does the trick. The uniqueness follows from the cancellation law for multiplication in R. (Details?) The rational number $u/v = b/a$ is obtained by taking $c = d = 1$.

q.e.d.

Definition 22(a). For each $a/b \in R$, $a \neq 0$, the unique $u/v \in R$ such that $a/b \circledast u/v = 1/1$ is the *reciprocal* or *multiplicative inverse* of a/b. It is denoted by

$$\left(\frac{a}{b}\right)^{-1}.$$

corollary 1: If $a/b \in R$, $a \neq 0$, then $(a/b)^{-1} = b/a$.

proof: Definition 19 and Definition 22(a).

q.e.d

Definition 22(b). For each a/b and each c/d in R, $a/b \neq 0/1$, the unique x/y such that $a/b \circledast x/y = c/d$ is the *quotient of c/d by a/b*. It is denoted by

$$\frac{c}{d} \div \frac{a}{b}.$$

corollary 2:   If $a/b \neq 0/1$ then

$$\frac{c}{d} \div \frac{a}{b} = \frac{cb}{da} = \frac{c}{d} \circledast \left(\frac{a}{b}\right)^{-1},$$

$$\frac{1}{1} \div \frac{a}{b} = \left(\frac{a}{b}\right)^{-1} = \frac{b}{a}.$$

proof: Theorem 25 and Definition 22.

q.e.d.∗

A familiar rule of elementary arithmetic is:

To divide one fraction by another, invert the second and multiply. If we regard b/a as the inversion of a/b then we see that this rule is established by Corollary 2.

The theory of order for the rational numbers is disposed of quickly and easily by a brief sequence of definitions and simple theorems.

∗ Definition 23.   (a) An element $a/b \in R$ is *positive* if and only if $ab > 0$.
  (b) An element $a/b \in R$ is *negative* if and only if $a/b \neq 0/1$ and a/b is not positive.
  (c) a/b *is greater than* c/d if and only if $a/b - c/d$ is positive; we write "$a/b > c/d$."
  (d) a/b *is less than* c/d if and only if c/d is greater than a/b.   If a/b is less than c/d we write "$a/b < c/d$."∗
  Note that if a/b is positive and if $a/b = a_1/b_1$, then a_1/b_1 is positive. (Why?)
  We emphasize that "$>$" and "$>$" (as well as "$<$" and "$<$") have different meanings.   The latter symbol stands for "is greater than" for the integers; it is used to define "positive" and "is greater than" for the rational numbers.

∗ Theorem 27.   (i) a/b is positive if and only if $a/b > 0/1$.
  (ii) Let (m,n) be the reduction of (a,b) to lowest terms.   Then a/b is positive if and only if m is a positive integer.

proof: (i) Since $a/b = a/b - 0/1$, it follows that a/b is positive if and only if $a/b - 0/1$ is positive.   Hence, by Definition 23(c), a/b is positive if and only if $a/b > 0/1$.
  (ii) Since $a/b = m/n$, $an = mb$ where n is a positive integer.   If a/b is positive, then $ab > 0$ and therefore a and b are both positive or both negative integers.   In either case, m must be a positive integer.   Con-

versely, if m is a positive integer, a and b are both positive or both negative integers. Hence, $ab > 0$ and therefore a/b is positive.

q.e.d.

Theorem 28. (i) a/b is negative if and only if $ab < 0$.

(ii) a/b is negative if and only if $-a/b$ is positive.

(iii) a/b is negative if and only if $0/1 > a/b$ (if and only if $a/b < 0/1$).

(iv) Let (m,n) be the reduction of (a,b) to lowest terms. Then a/b is negative if and only if m is a negative integer.

proof: (i) Note, first, that $a/b = 0/1$ if and only if $ab = 0$. Now, suppose a/b is negative. Then $ab \neq 0$ and $ab \not> 0$. By the order properties of the integers, $ab < 0$. Conversely, if $ab < 0$ then $ab \neq 0$ and $ab \not> 0$. Thus $a/b \neq 0/1$ and a/b is not positive. By Definition 23(b), a/b is negative.

(ii) By (i), a/b is negative if and only if $ab < 0$. By the order properties of the integers, $ab < 0$ if and only if $(-a)b > 0$. Therefore, a/b is negative if and only if $-a/b \ (= a/-b = -(a/b))$ is positive.

(iii) Since $-a/b = -(a/b) = 0/1 - a/b$, it follows that $0/1 - a/b$ is positive if and only if $-a/b$ is positive. Therefore, by (ii), $0/1 - a/b$ is positive, i.e., $0/1 > a/b$, if and only if a/b is negative. By Definition 23(d), a/b is negative if and only if $a/b < 0/1$.

(iv) Exercise.

q.e.d.

Theorem 29. (i) If a/b, c/d are rational numbers, then one and only one of

$$\frac{a}{b} < \frac{c}{d}, \quad \frac{a}{b} = \frac{c}{d}, \quad \frac{c}{d} < \frac{a}{b},$$

holds. (Trichotomy law of $<$).

(ii) If $a/b < c/d$ and $c/d < e/f$ then $a/b < e/f$. (Transitivity of $<$).

proof: Exercise.

Definition 24. $a/b \leq c/d$ if and only if $a/b < c/d$ or $a/b = c/d$. $a/b \geq c/d$ if and only if $a/b > c/d$ or $a/b = c/d$.

Definition 25. $<$ and $>$ are *strict inequalities*; $\leq$ and $\geq$ are *weak inequalities*.∗

EXERCISES

Prove:

1. $a/1 > b/1$ if and only if $a > b$; $a/1 = b/1$ if and only if $a = b$.

2. One and only one of: $a/b > c/d$, $a/b = c/d$, $c/d > a/b$ holds.

3. If $a/b > c/d$ and $c/d > e/f$ then $a/b > e/f$.

4. If $a/b \le c/d$ and $c/d < e/f$ then $a/b < e/f$.

5. If $a/b = c/d$ and $c/d \le e/f$ then $a/b \le e/f$.

6. If $a/b > c/d$ then for each $e/f \in R$, $a/b \perp e/f > c/d \perp e/f$.

7. If $a/b > c/d$ and e/f is positive, then $a/b \circledast e/f > c/d \circledast e/f$; if e/f is negative, then $a/b \circledast e/f < c/d \circledast e/f$.

8. A sum of positive rational numbers is positive (use induction).

9. If a/b and c/d are both positive or both negative then $a/b \circledast c/d$ is positive; otherwise, $a/b \circledast c/d$ is negative.

10. A sum of squares of rational numbers is nonnegative. (Define $(a/b)^2 = a/b \circledast a/b$ and use induction.) Hence, $1/1$ is positive.

Our final task in developing the arithmetic of R is to show that R contains a "copy" of the integers Z. The proof that it does requires the concept of isomorphism (see footnote 2), page 143).

* Definition 26. A subset A of R is isomorphic with Z if and only if there is a one-one correspondence g between Z and A such that:
For all $x, y \in Z$,

1. $g(x + y) = g(x) \perp g(y)$;
2. $g(xy) = g(x) \circledast g(y)$; and
3. if $x > y$ then $g(x) > g(y)$.

The mapping g is an *isomorphism* between Z and A.

Now let Z^* be the subset of R defined by

$$Z^* = \left\{ \frac{a}{1} \,\middle|\, a \in Z \right\}.$$

Note that a rational number c/d is an element in Z^* if and only if $d \mid c$. Indeed, if $d \mid c$, then there is an $x \in Z$ such that $c = xd$. Hence, $c/d = xd/d = x/1$, whence $c/d \in Z^*$. Conversely, if $c/d \in Z^*$, then there is a $y \in Z$ such that $c/d = y/1$. Therefore, $c = c \cdot 1 = y \cdot d$ and $d \mid c$.

The theorem we wish to prove is

Theorem 30. Z^* is isomorphic with Z.

proof: To prove the theorem, we exhibit an isomorphism g between Z and Z^*. Define g by

$$g(a) = \frac{a}{1}, \quad a \in Z.$$

By definition, $\mathfrak{D}(g) = Z$ and $\mathfrak{R}(g) = Z^*$. Since $a/1 = b/1$ if and only if $a = b$, it follows that g is a one-one correspondence between Z and Z^*. Further:

1. $g(a + b) = (a + b)/1 = a/1 \perp b/1 = g(a) \perp g(b)$, (definition of $\perp$);
2. $g(ab) = ab/1 = a/1 \circledast b/1 = g(a) \circledast g(b)$, (definition of $\circledast$);
3. $a/1 > b/1$ if and only if $a > b$, hence $g(a) > g(b)$ if and only if $a > b$.

By Definition 26, g is an isomorphism between Z and Z^*.

q.e.d.

We have proved before (Theorem 10) that there is an isomorphism f between N, the set of natural numbers, and the set $\mathfrak{N}$ of nonnegative integers, where

$$\mathfrak{N} = \{x \mid x \in Z \text{ and } x \geq 0\}.$$

Let

$$\mathfrak{N}^* = \{a/1 \mid a \in Z \text{ and } a \geq 0\}.$$

Thus $\mathfrak{N}^*$ is the set of nonnegative elements in Z^*. Further, let g' be the restriction of g to $\mathfrak{N}$. It is easy to see that g' is an isomorphism between $\mathfrak{N}$ and $\mathfrak{N}^*$. But now one can verify, with very little trouble, that the composite $g' \circ f$ is a one-one correspondence between N and $\mathfrak{N}^*$ such that for all $m, n \in N$,

$$(g' \circ f)(m + n) = (g' \circ f)(m) \perp (g' \circ f)(n),$$
$$(g' \circ f)(mn) = (g' \circ f)(m) \circledast (g' \circ f)(n),$$

and

$$(g' \circ f)(m) > (g' \circ f)(n) \text{ if and only if } m > n;$$

it is therefore reasonable to call $g' \circ f$ an isomorphism between N and $\mathfrak{N}^*$. Hence,

Theorem 31. There exists an isomorphism between N and $\mathfrak{N}^*$.

In elementary arithmetic, we are accustomed to regarding a rational number, say $\frac{5}{1}$, as being the same as the integer, 5. We shall want to do the same kind of thing here. Of course, we cannot accept, literally, the statement

(3.17)
$$\frac{5}{1} = 5.$$

The integer 5 is an equivalence class of ordered pairs of natural numbers; thus, 5 is a subset of $N \times N$. The rational number $\frac{5}{1}$, on the other hand, is an equivalence class of $Z \times Z_0$ where $Z = N \times N$ and Z_0 is the subset of $N \times N$ consisting of all ordered pairs (a,b), a, $b \in N$, where $a \neq b$. Such being the case, it is not true that $\frac{5}{1}$ and 5 are names for the same set. Consequently, equation (3.17) is false. However, by using the isomorphism g defined in Theorem 30, we can introduce the desired convention via the back door. Since for each $a/1 \in Z^*$ there is one and only one $a \in Z$ such that

$$g(a) = \frac{a}{1},$$

we shall agree that henceforth the symbol,

$$a,$$

will denote the rational number $a/1$. Consequently, from now on, the symbols,

$$5, \quad 17, \quad -8, \quad -3, \quad 24,$$

will stand for the rational numbers,

$$\frac{5}{1}, \quad \frac{17}{1}, \quad \frac{-8}{1}, \quad \frac{-3}{1}, \quad \frac{24}{1},$$

respectively. If, in any discourse, the symbol "5" appears, it must be stated explicitly (or should be understood from the context) whether a natural number, an integer, or the rational number $\frac{5}{1}$ is intended. A symbol such as "-5" can only represent an integer or a rational number; a symbol such as "$\frac{5}{3}$" can only denote a rational number. In most cases, where the various systems are being used for computational purposes, it does not matter whether we think of 5 as being a natural number, an integer, or a rational number. It is only when we are concerned with the nature of numbers that such distinctions are important.

The symbols ">" and "<" have also served their function, that of distinguishing between the order relations for the rationals and the order relations for the integers. From now on, we agree that:

">" denotes *is greater than* for the natural numbers, the integers *and* the rationals; "<" denotes *is less than* for the same three number systems.

Let us observe that with our new conventions,

$$a \div b = \frac{a}{1} \div \frac{b}{1} = \frac{a}{b},$$

so that we may think of the rational number a/b as "a divided by b."

EXERCISES

1. Let m and n be positive integers such that $m < n$, and let $r \in R$. Prove:
(a) If $1 < r$ then $r^m < r^n$.
(b) If $0 < r < 1$ then $r^n < r^m$.

2. Let r and s be positive rational numbers. Prove that there is an integer n such that $nr > s$. (This result is known as the *Archimedean Law* for the rational numbers).

The concept of the "absolute value function" for the rational numbers will be important in Chapters 4 and 5.

* Definition 27. The *absolute value function* on R is the set

$$|| = \{(x,x) \mid x \in R \text{ and } x \geq 0\} \cup \{(x,-x) \mid x \in R \text{ and } x < 0\}.$$

In the customary functional notation, this function is defined by

$$|x| = \begin{cases} x, \text{ if } x \in R \text{ and } x \geq 0, \\ -x, \text{ if } x \in R \text{ and } x < 0. \end{cases}$$

We leave it to the reader to verify that properties 1–8 (page 149) also hold for all rational numbers.*

3.6.   CONCLUSION: INTEGRAL DOMAINS AND QUOTIENT FIELDS[3]

The procedure by which we constructed the rationals from the integers has a more general utility than is evident from Section 3.4.

In constructing the rationals R from the integers Z, not all of the properties of Z were used; indeed, not all of the properties of Z are known. An examination of the method followed in Section 3.4 will reveal that the only properties of Z which played a role in this construction were:

For each a, b, $c \in Z$,

1. $a + b = b + a$ (commutative law for $+$);
2. $a + (b + c) = (a + b) + c$ (associative law for $+$);
3. there is a *zero element*, $0 \in Z$, such that $a + 0 = a$;
4. there exists $a_1 \in Z$ such that $a + a_1 = 0$; a_1 is the *additive inverse* of a;
5. $ab = ba$ (commutative law for $\cdot$);
6. $a(bc) = (ab)c$ (associative law for $\cdot$);
7. there is a *unity element* $1 \in Z$, $1 \neq 0$, which has the property $a \cdot 1 = a$;
8. $a(b + c) = ab + ac$ (multiplication is distributive over addition);
9. if $ab = ac$ and $a \neq 0$, then $b = c$ (cancellation law for multiplication).

The relation $<$ on Z was used in order to define a like relation on R, but it served no purpose in constructing the elements of R or in defining the binary operations on R. After its construction, it turned out that, with respect to its two binary operations, R had not only properties 1–9 but also a tenth one; namely,

10. For all a, $b \in R$, $a \neq 0$, the equation $ax = b$ has a unique solution x.

The system Z does not have property 10.

To facilitate the discussion, we introduce two new terms.

* Definition 28.   A set I on which are defined two binary operations, $+$ and $\cdot$, satisfying conditions 1–9 (above) is an *integral domain*.*

---

[3] Only Definitions 28 and 29 of this concluding section are required for a reading of Chapter 4.

As a consequence of conditions 3 and 7, an integral domain contains at least two elements.

* **Definition 29.** A *field* is an integral domain which also satisfies condition 10.*

So far we know that Z and R are both integral domains, R is a field, but Z is not. In Chapter 4 we shall construct the real numbers R and the complex numbers C and prove that R and C are fields. Taking these results for granted, we ask: are there systems, other than Z, satisfying 1–9 but not 10? On the basis of a few assumptions which will be justified in Chapters 4 and 5, it is possible to exhibit such systems so that the question is answered affirmatively.

We assume first that there is a number, $\sqrt{2}$, i.e., a number $a > 0$ such that $a^2 = 2$, and that $\sqrt{2} \notin R$. Next, we assume we know what is meant by $a + b\sqrt{2}$ [4] where a and b are integers; these assumptions correspond to things we have done in the past and should not arouse too great a revulsion. Set

$$Z[\sqrt{2}] = \{a + b\sqrt{2} \mid a, b \in Z\}$$

and define

$$(a + b\sqrt{2}) \oplus (c + d\sqrt{2}) = (a + c) + (b + d)\sqrt{2},$$
$$(a + b\sqrt{2}) \odot (c + d\sqrt{2}) = (ac + 2bd) + (ad + bc)\sqrt{2}.$$

It is a simple task to verify that the $\oplus$ and $\odot$ so defined are binary operations on $Z[\sqrt{2}]$, that conditions 1–9 are satisfied, and therefore $Z[\sqrt{2}]$ is an integral domain. Moreover, it is easy to see that condition 10 is not satisfied, hence $Z[\sqrt{2}]$ is not a field.

EXERCISES

1. Carry out a few of the verifications required to show that $Z[\sqrt{2}]$ is an integral domain but not a field.

2. Making the proper assumptions concerning $\sqrt{3}$ and $a + b\sqrt{3}$, $a, b \in Z$, show that $Z[\sqrt{3}] = \{a + b\sqrt{3} \mid a, b \in Z\}$ is an integral domain but not a field.

---

[4] This will be clarified in Chapters 4 and 5.

3. What can you say about $\{a + b\sqrt{-1} \mid a, b \in Z\}$?
$\{2a + b\sqrt{3} \mid a, b \in Z\}$? $\{a + b\sqrt{4} \mid a, b \in Z\}$?

4. Prove that an integral domain has only one zero element and only one unity element. By Definition 29 this statement is true, also, for all fields.

5. Let G be a set on which are defined two binary operations $+$ and $\cdot$, satisfying conditions 1–8 and 10 of Definitions 28 and 29. Prove that condition 9 is deducible from the other conditions.

From the examples, one can infer that there are many integral domains which are not fields. In fact, there are integral domains, which are not fields, bearing very little resemblance (other than properties 1–9) to the examples above.

We now come to the point of this section. Since the field of rationals was constructed from the integers by using only properties 1–9 of the integers, it is certainly plausible that, given any integral domain I, it is possible to construct a field from I by the method of Section 3.4. The resulting field Q is the *quotient field* of I.

Theorem 32. For every integral domain I there exists a quotient field Q of I.

We give a step-by-step outline of the proof, noting the corresponding definitions and theorems of Section 3.4. In carrying out the details of the proof of Theorem 32, one need only replace the word "integer" by "element of I" and "rational number" by "element of Q." It will be helpful to use Greek letters to denote the elements of I in order to emphasize the fact that the proof of Theorem 32 depends only on properties 1–9.

   1. **Definition.** $I_0 = I - \{0\}$. (Definition 16.)
   2. **Definition.** For all (α,β), $(\gamma,\delta) \in I \times I_0$, define $(\alpha,\beta) \sim (\gamma,\delta)$ if and only if $\alpha\delta = \gamma\beta$. (Definition 17.)
   3. **Theorem.** $\sim$ is an equivalence relation. (Theorem 17.)
   4. **Definition.** $< \alpha,\beta > = \{(\gamma,\delta) \mid (\alpha,\beta) \sim (\gamma,\delta)\}$. Q is the set of all equivalence classes $< \alpha,\beta >$. (Definition 18.) (In studying the rational numbers we wrote "a/b" in place of "$< a,b >$".)
   5. **Theorem.** $< \alpha,\beta > = < \gamma,\delta >$ if and only if $\alpha\delta = \gamma\beta$. (Theorem 18.)

6. **Theorem.** If $< \alpha,\beta > = < \gamma,\delta >$ and $< \mu,\nu > = < \sigma,\tau >$ then

$$< \alpha\nu + \mu\beta, \beta\nu > = < \nu\tau + \sigma\delta, \delta\tau >$$

and

$$< \alpha\mu,\beta\nu > = < \gamma\sigma,\delta\tau >. \quad \text{(Theorem 19.)}$$

7. **Corollary.** $< \alpha,\beta > \perp < \mu,\nu > = < \alpha\nu + \mu\beta, \beta\nu >$

and

$$< \alpha,\beta > \circledast < \mu,\nu > = < \alpha\mu,\beta\nu >.$$

$(< \alpha,\beta >, < \mu,\nu > \in Q)$ are binary operations on Q. (Corollary to Theorem 19.)

8. **Definition.** $\perp$ and $\circledast$ are *sum* and *product*, respectively, of elements of Q. (Definition 19.)

9. **Theorem.** $< 0,1 >$ is a zero element for Q. (Theorem 22.)

10. **Theorem.** $\perp$ and $\circledast$ are commutative and associative; $\circledast$ is distributive over $\perp$. Cancellation laws hold for $\perp$ and $\circledast$. (Theorem 23.)

11. **Theorem.** For all $< \alpha,\beta >$, $< \gamma,\delta >$ there exists an element $< \mu,\nu >$ such that $< \alpha,\beta > = < \mu,\nu > \perp < \gamma,\delta >$. In particular, there exists an element $< \mu,\nu >$ such that $< 0,1 > = < \mu,\nu > \perp < \gamma,\delta >$ and this element is an *additive inverse* of $< \gamma,\delta >$. (Theorem 24.)

12. **Theorem.** $< 1,1 >$ is a unity element for Q. (Theorem 25.)

13. **Theorem.** For each $< \alpha,\beta > \neq < 0,1 >$ and for each $< \gamma,\delta >$ there exists an element $< \mu,\nu >$ such that $< \alpha,\beta > \circledast < \mu,\nu > = < \gamma,\delta >$. (Theorem 26.)

EXERCISES

1. Construct the quotient fields of $Z[\sqrt{2}]$ and $Z[\sqrt{-1}]$. Are there simple ways of representing the elements of these fields?

2. Suppose I is already a field; what can you say about the quotient field of I? (Hint: Define the concept of isomorphism for fields and prove that if I is a field then I is isomorphic to its quotient field.)

3. Discuss the generalized operations in an arbitrary integral domain I.

4

THE REAL NUMBERS

4.1. THE MYSTERIOUS $\sqrt{2}$

In elementary geometry the following statement occurs: If each of the legs of a right isosceles triangle is one unit in length then the length of the hypotenuse is $\sqrt{2}$ units. A critical analysis of this statement raises a number of questions. One of them is, "What is meant by the length of a line segment?"; another is, "What is $\sqrt{2}$?". We address ourselves to the latter question. The usual answer is that the square root of 2 is the positive number whose square is 2; i.e., $\sqrt{2}$ is the positive number which when multiplied by itself gives the result 2. But the answer leads immediately to another question, "What kind of number?" The preceding work has taught us that there are several kinds of numbers; there are the natural numbers, the integers, the rational numbers. So, if one says that $\sqrt{2}$ is a number, it should be specified what kind of number one has in mind. Now certainly there is no natural number or integer whose square is 2. But it is also true, and this fact came as a great shock to mathematicians when it was discovered over two millennia ago, that there is no rational number whose square is the rational number 2. It is worth digressing to prove this fact. A few preliminaries are needed.

* Definition 1. Let n be an integer. n is *even* if and only if there is an integer q such that $n = 2q$. The integer n is *odd* if and only if there is an integer q such that $n = 2q + 1$.

Theorem 1. For all integers m and n,

    (a) n is even or n is odd, but not both;

    (b) $m + n$ is even if and only if both m and n are even or both m and n are odd;

    (c) $m \cdot n$ is even if and only if at least one of m and n is even.

proof: Exercise.

Theorem 2. There is no rational number whose square is 2.

proof: Suppose $\alpha \in R$ and $\alpha^2 = 2$. Then $\alpha = p/q$ where p and q are integers relatively prime to each other. In particular p and q are not both even, since they would then have the common factor 2. We shall now show, however, that p and q must both be even. This contradiction will prove the theorem. Indeed, since $\alpha^2 = 2$, it follows that $p^2/q^2 = 2$, whence $p^2 = 2q^2$. Now $2q^2$ is even, hence p^2 is even, and hence p is even (for if p were odd, $p^2 = p \cdot p$ would be odd). This means p is an integral multiple of 2, say $p = 2r$, where r is an integer. Then $2q^2 = p^2 = 4r^2$, and dividing by 2, $q^2 = 2r^2$. But $2r^2$ is even, hence q^2 is even, and so q is even. Thus both p and q are even.

q.e.d._{*}

So, if we say that $\sqrt{2}$ is a number whose square is 2, we cannot mean $\sqrt{2}$ is a rational number; we must mean $\sqrt{2}$ is some other kind of number. It would seem, then, that there is some type of number system which is an extension of the rational number system (or, more precisely, contains an arithmetic copy of the rationals) and in which there is a number whose square is 2. In fact, there is such a number system, the so-called real numbers. In this system there is a $\sqrt{2}$ and a great many things can be done which are impossible in the rationals.

To motivate what follows we are going to return to geometrical ideas. It is not our purpose to go into the foundations of geometry at this point, and therefore what we say about geometrical concepts is to be taken, not as precise mathematical statements but as heuristic guides. Our ultimate definition of real number will—in the spirit of this book—be stated entirely within the framework of set theory, and will be precise.

To return to where we started, $\sqrt{2}$ arose in expressing a certain length, or in other words in expressing the distance between two points. Now in practical day-to-day life how does one go about expressing the distance between two points? One measures it, either very roughly or with much precision. In any case, a certain amount of error is to be expected no matter how elaborate a technique of measurement is devised. The result of the measurement is invariably expressed as a certain rational number of units of length. Let us fix the unit here as the foot. Then we speak of a distance measured as $1\frac{1}{2}$ ($= \frac{3}{2}$) feet, or 3.6245 ($= \frac{36245}{10000}$) feet, or 2 feet 7 inches ($= \frac{31}{12}$ feet). As we said before, in any measurement a certain amount of error is to be expected. In practice this means that if a distance is measured with, say, a yardstick, and if the result of the measurement is $1\frac{1}{2}$ feet, it is quite possible that a more precise measuring procedure would yield, say, 1.502 feet, and that a still more precise technique would yield 1.501873 feet as the distance, and so on. This suggests that in order to obtain a distance[1] we should embark upon an unending sequence of measurements, each more precise than its predecessors, and that successive measurements will in turn approach more closely the value of the distance.

*Definition 2. A *sequence of rational numbers* is a mapping $\varphi \colon N_1 \longrightarrow R$.

For any $n \in N_1$, $\varphi(n)$ is the nth *term of the sequence.**

Thus, if $\varphi \colon N_1 \longrightarrow R$ is a sequence, $\varphi(1)$ is the first term of the sequence, $\varphi(2)$ is the second, and so on. A sequence of rational numbers is clearly an appropriate mathematical concept to use in speaking of an unending series of measurements; namely, the result of the nth measurement is $\varphi(n)$ feet. Now the results of series of measurements which are conducted with ever-increasing precision will give rise to rather special types of sequences. A sequence such as $\varphi \colon N_1 \longrightarrow R$ with $\varphi(1) = 2$, $\varphi(2) = 17$, $\varphi(3) = 1$, $\varphi(4) = 8\frac{13}{17}$, $\varphi(5) = 2{,}743{,}852\frac{1}{2}$, etc., which continues perpetually to fluctuate wildly will not occur. A sequence which arises from a series of increasingly exact measurements is distinguished by the following: as one makes more and more measurements, *the results of later measurements will all be quite close to one another.* The meanings of "later" and "quite close" as applied to sequences is vague. These terms will be made precise below. But before proceeding to a study of sequences having the property that the "later" terms are

---

[1] We are here taking the naive point of view that we can determine completely the location of "physical points" and that we know what is meant by the distance between two such points.

"quite close" to each other—the so-called Cantor sequences—we need some arithmetic tools for the handling of sequences.

4.2. THE ARITHMETIC OF SEQUENCES

*In the remainder of this chapter "$\mathcal{S}$" will denote the set of all sequences (of rational numbers). As a preliminary to Definitions 3–6 below, observe the following:

If σ_1 and σ_2 are sequences, then

$$\sigma_1(n) + \sigma_2(n), \ n \in N_1$$

is also a sequence. Similarly,

$$\sigma_1(n) - \sigma_2(n), \ n \in N_1$$

and

$$\sigma_1(n) \cdot \sigma_2(n), \ n \in N_1$$

and

$$-(\sigma_1(n)), \ n \in N_1$$

are all sequences. These results are established easily and are left as exercises.

Definition 3. If σ_1 and σ_2 are sequences, the *sum*, $\sigma_1 + \sigma_2$, of σ_1 and σ_2, is the sequence defined by

$$(\sigma_1 + \sigma_2)(n) = \sigma_1(n) + \sigma_2(n), \ n \in N_1.*^2$$

EXAMPLE

$\sigma_1(n) = (\tfrac{1}{2})^n, \ n \in N_1, \ \sigma_2(n) = (-\tfrac{1}{2})^n, \ n \in N_1.$ Then

$$(\sigma_1 + \sigma_2)(n) = \begin{cases} 0, & \text{if } n \text{ is odd,} \\ \dfrac{1}{2^{n-1}}, & \text{if } n \text{ is even.} \end{cases}$$

---

[2] *A word of warning*: The symbol "$+$" occurring in the definition of sum of sequences has two meanings. On the right side of the defining equation occurs "$\sigma_1(n) + \sigma_2(n)$" denoting a sum of rational numbers, and this has been defined in Chapter 3. On the left side of the equation there is "$\sigma_1 + \sigma_2$" and this symbol is being defined in terms of sums of rational numbers. It is regrettable that a few symbols have to bear so many meanings. This could be a source of confusion. However, the alternative would require the use of so many symbols that possibly even greater confusion would result. We hope that this warning will enable the reader to understand the various meanings assigned to "$+$" and "$\cdot$" with a minimum of inconvenience.

* Definition 4. If σ is a sequence, then $-\sigma$ is the sequence defined by

$$(-\sigma)(n) = -(\sigma(n)),\ n \in N_1.*$$

EXAMPLE

If $\sigma(n) = 1/2^n$, $n \in N_1$, then $(-\sigma)(n) = -1/2^n$, $n \in N_1$.

* Definition 5. If σ_1, σ_2 are sequences, then $\sigma_1 - \sigma_2$ is the sequence defined by

$$\sigma_1 - \sigma_2 = \sigma_1 + (-\sigma_2).*$$

Thus,

$$
\begin{aligned}
(\sigma_1 - \sigma_2)(n) &= (\sigma_1 + (-\sigma_2))(n) = \sigma_1(n) + (-\sigma_2)(n) \\
&= \sigma_1(n) + (-(\sigma_2(n))) \\
&= \sigma_1(n) - \sigma_2(n),\ \text{for all}\ n \in N_1,
\end{aligned}
$$

where the last "$-$" denotes subtraction in R.

* Definition 6. If σ_1, σ_2 are sequences, the *product of σ_1 and σ_2, $\sigma_1 \cdot \sigma_2$,* is the sequence defined by

$$(\sigma_1 \cdot \sigma_2)(n) = \sigma_1(n) \cdot \sigma_2(n),\ n \in N_1.*$$

EXAMPLE

If $\sigma_1(n) = 1/2^n$, $n \in N_1$, $\sigma_2(n) = 2^n$, $n \in N_1$, then $(\sigma_1 \cdot \sigma_2)(n) = 1$ for all $n \in N_1$.

These operations on sequences correspond quite naturally to the analogous operations on R. The relationship is often expressed by saying that the operations on sequences are defined term by term from the operations on R. These definitions have a natural connection with the theory of measurement. For instance, suppose A, B and C are three points on a straight line with B between A and C. A common method of measuring the distance from A to C is to measure separately the distances from A to B and from B to C and to add the results. To obtain a sequence of measurements from A to C one can thus add term by term the results of a sequence of measurements of the distance from A to B, and a sequence of measurements of the distance from B to C.

In a like fashion, to obtain a sequence of measurements for the area of a rectangle one could multiply the results of measuring the length and breadth of the rectangle.

* Definition 7. Let $r \in R$; the sequence $\hat{r}$ defined by $\hat{r}(n) = r$ for all $n \in N_1$, is a *constant sequence.**

EXAMPLES

1. The constant sequences $\hat{0}$ and $\hat{1}$ defined, respectively, by $\hat{0}(n) = 0, n \in N_1$, and $\hat{1}(n) = 1, n \in N_1$, are especially important.

2. $\widehat{\frac{15}{7}}$ is the sequence defined by $\widehat{\frac{15}{7}}(n) = \frac{15}{7}, n \in N_1$.

EXERCISES

1. Let $\sigma_1(n) = 1 - 1/2^n, n \in N_1$, $\sigma_2(n) = 1 + 1/2^n, n \in N_1$. Find $-\sigma_1, -\sigma_2, \sigma_1 + \sigma_2, \sigma_1 - \sigma_2, \sigma_2 - \sigma_1$ and $\sigma_1 \cdot \sigma_2$.

2. Do the same for the sequences $\sigma_1(n) = 1/3^n, n \in N_1$; $\sigma_2(n) = 3^n$, $n \in N_1$.

* Theorem 3. For all $\sigma, \sigma_1, \sigma_2, \sigma_3 \in S$:

1. $(\sigma_1 + \sigma_2) + \sigma_3 = \sigma_1 + (\sigma_2 + \sigma_3)$;
2. $\sigma_1 + \sigma_2 = \sigma_2 + \sigma_1$;
3. $\sigma + \hat{0} = \sigma$;
4. $\sigma + (-\sigma) = \hat{0}$;
5. $(\sigma_1 \cdot \sigma_2) \cdot \sigma_3 = \sigma_1 \cdot (\sigma_2 \cdot \sigma_3)$;
6. $\sigma_1 \cdot \sigma_2 = \sigma_2 \cdot \sigma_1$;
7. $\sigma \cdot \hat{1} = \sigma$;
8. $\sigma_1 \cdot (\sigma_2 + \sigma_3) = \sigma_1 \cdot \sigma_2 + \sigma_1 \cdot \sigma_3$.

proof: The proof is a matter of straightforward verifications based upon the following idea:

Suppose f and g are functions having the same domain $\mathfrak{D}$. To prove $f = g$ one shows that $f(x) = g(x)$ for each $x \in \mathfrak{D}$.

We illustrate by proving parts 1, 3, 7 and 8; the proofs of the remaining statements are left to the reader.

1. To prove $(\sigma_1 + \sigma_2) + \sigma_3 = \sigma_1 + (\sigma_2 + \sigma_3)$ we show that for each $n \in N_1$, $((\sigma_1 + \sigma_2) + \sigma_3)(n) = (\sigma_1 + (\sigma_2 + \sigma_3))(n)$. For each $n \in N_1$ we have

$$((\sigma_1 + \sigma_2) + \sigma_3)(n) = (\sigma_1 + \sigma_2)(n) + \sigma_3(n) \qquad \text{(Definition of sum of sequences)}$$

$$= (\sigma_1(n) + \sigma_2(n)) + \sigma_3(n) \qquad \text{(Definition of sum of sequences)}$$

$$= \sigma_1(n) + (\sigma_2(n) + \sigma_3(n)) \qquad \text{(Associativity of sum in } R)$$

$$= \sigma_1(n) + (\sigma_2 + \sigma_3)(n) \qquad \text{(Definition of sum of sequences)}$$

$$= (\sigma_1 + (\sigma_2 + \sigma_3))(n) \qquad \text{(Definition of sum of sequences)}$$

Hence, 1 follows.

3. $(\sigma + \hat{0})(n) = \sigma(n) + \hat{0}(n) = \sigma(n) + 0 = \sigma(n)$, for each $n \in N_1$. Therefore, $\sigma + \hat{0} = \sigma$.

7. $(\sigma \cdot \hat{1})(n) = \sigma(n) \cdot \hat{1}(n) = \sigma(n) \cdot 1 = \sigma(n)$, for each $n \in N_1$, whence $\sigma \cdot \hat{1} = \sigma$.

8. For each $n \in N_1$,

$$(\sigma_1 \cdot (\sigma_2 + \sigma_3))(n) = \sigma_1(n) \cdot (\sigma_2 + \sigma_3)(n) \qquad \text{(Definition of product of sequences)}$$

$$= \sigma_1(n) \cdot (\sigma_2(n) + \sigma_3(n)) \qquad \text{(Definition of sum of sequences)}$$

$$= \sigma_1(n) \cdot \sigma_2(n) + \sigma_1(n) \cdot \sigma_3(n) \qquad \text{(Distributive law in } R)$$

$$= (\sigma_1 \cdot \sigma_2)(n) + (\sigma_1 \cdot \sigma_3)(n) \qquad \text{(Definition of product of sequences)}$$

$$= ((\sigma_1 \cdot \sigma_2) + (\sigma_1 \cdot \sigma_3))(n), \qquad \text{(Definition of sum of sequences)}$$

therefore, $\sigma_1 \cdot (\sigma_2 + \sigma_3) = (\sigma_1 \cdot \sigma_2) + (\sigma_1 \cdot \sigma_3)$.

q.e.d.∗

Readers who are familiar with some of the concepts of abstract algebra will recognize that $\mathcal{S}$, with the binary operations $+$ and $\cdot$, is a commutative ring with unity element. It is not an integral domain (see Definition 28, Chapter 3) since, as the following example shows, the cancellation law for multiplication of sequences does not hold:

Let the sequences σ_1, σ_2, σ_3 be defined by

$$\sigma_1(1) = 1,\ \sigma_1(n) = 0 \text{ for all } n > 1;$$
$$\sigma_2(1) = 0,\ \sigma_2(n) = 1 \text{ for all } n > 1;$$
$$\sigma_3(1) = 0,\ \sigma_3(2) = 1,\ \sigma_3(n) = 0 \text{ for all } n > 2,$$

respectively. Then $\sigma_1 \cdot \sigma_2 = \sigma_1 \cdot \sigma_3 = \hat{0}$ and $\sigma_1 \neq \hat{0}$ but $\sigma_2 \neq \sigma_3$.

Any sequence of numbers which arises from more and more precise measurements of a physical object is unlikely to have ever larger numbers in it. For instance, one would not expect a sequence like

$$\sigma(n) = \begin{cases} 2^n, & n \text{ even,} \\ 1 + \dfrac{1}{2^n}, & n \text{ odd} \end{cases}$$

to result from a series of increasingly refined measurements of a given object. On the contrary, it is far more reasonable to believe that all the numbers obtained by such measurements would be below some "bound."

* Definition 8. A sequence β is *bounded* if and only if there exists a positive rational number r such that $|\beta(n)| \leq r$ for all $n \in N_1$. Any positive rational number r such that $|\beta(n)| \leq r$ is a *bound* for β. The set of all bounded sequences is denoted by "$\mathcal{B}$".*

EXERCISES

1. Prove that every constant sequence is bounded. Hence, in particular, $\hat{0}$ and $\hat{1} \in \mathcal{B}$.

2. Prove: If r is a bound for a sequence β then every rational number $s > r$ is also a bound for β.

3. Find a bound for the sequence

$$\beta(n) = \begin{cases} 2 - \dfrac{1}{2^n}, & \text{for } n \text{ odd,} \\ 1 + \dfrac{1}{2^n}, & \text{for } n \text{ even.} \end{cases}$$

4. Find a bound for the sequence

$$\beta(n) = \begin{cases} 1, & \text{for } n \text{ odd,} \\ -1, & \text{for } n \text{ even.} \end{cases}$$

5. For each $n \in N_1$, define $s_n = \sum\limits_{j=1}^{n} \dfrac{1}{j}$ and set

$$\beta(n) = s_n, \ n \in N_1.$$

Is β a bounded sequence? Prove it. (Consult the chapter on series in any calculus book.)

The arithmetic of the set $\mathcal{B}$ is described by

* Theorem 4. Sums, differences and products of bounded sequences are bounded sequences. Moreover, since $\hat{0}$ and $\hat{1}$ are elements of $\mathcal{B}$, properties 1–8 of Theorem 3 hold for $\mathcal{B}$ (hence $\mathcal{B}$ is a commutative ring with unity element).

proof: Let $\beta_1, \beta_2 \in \mathcal{B}$; then there exist positive rational numbers b_1, b_2 such that $|\beta_1(n)| \leq b_1$, $|\beta_2(n)| \leq b_2$ for all $n \in N_1$. Then $|(\beta_1 + \beta_2)(n)| = |\beta_1(n) + \beta_2(n)| \leq |\beta_1(n)| + |\beta_2(n)| \leq b_1 + b_2$, for all $n \in N_1$, therefore, $b_1 + b_2$ is a bound for $\beta_1 + \beta_2$, hence $\beta_1 + \beta_2$ is bounded.

The boundedness of $\beta_1 - \beta_2$ is clear. For, since β_2 is bounded so is $-\beta_2$ (why?). Hence, by the first part of the theorem so is $\beta_1 + (-\beta_2) = \beta_1 - \beta_2$.

For the product $\beta_1 \cdot \beta_2$ we have

$$
\begin{aligned}
|(\beta_1 \cdot \beta_2)(n)| &= |\beta_1(n) \cdot \beta_2(n)| \\
&= |\beta_1(n)| \cdot |\beta_2(n)| \leq b_1 \cdot b_2
\end{aligned}
$$

for all $n \in N_1$, whence $\beta_1 \cdot \beta_2$ is bounded and has $b_1 \cdot b_2$ as a bound.

That properties 1–8 of Theorem 3 hold for the sequences in $\mathcal{B}$ is clear. Indeed, since $\mathcal{B} \subset \mathcal{S}$ and $\hat{0}$, $\hat{1} \in \mathcal{B}$ and 1–8 hold for all the sequences in $\mathcal{S}$, they automatically hold for those in $\mathcal{B}$. (Therefore, $\mathcal{B}$ is a commutative ring with unity element.)

q.e.d.*

One says that "$\mathcal{B}$ is a subring of $\mathcal{S}$." Our development of arithmetic does not require the use of the concept of "subring" and therefore we do not give a definition of this term here. A definition as well as criteria for a subset of a ring to be a subring will appear in Book II, Algebra.

EXERCISES

1. Prove that $\mathcal{B}$ fails to satisfy the cancellation law for multiplication (hence is not an integral domain.)

2. If $\hat{r}$ is a constant sequence and β is a bounded sequence with bound b prove: (i) $b + | r |$ is a bound for $\beta + \hat{r}$ and (ii) $b \cdot | r |$ is a bound for $\beta \cdot \hat{r}$.

3. Let β_1 be the sequence defined in Exercise 3, page 184, β_2 the sequence defined in Exercise 4 of the same page. Find bounds for $\beta_1 + \beta_2$, $\beta_1 - \beta_2$, $\beta_2 - \beta_1$ and $\beta_1 \cdot \beta_2$.

Consider the sequence σ defined by

$$\sigma(1) = 10^{23},$$
$$\sigma(2) = -10^{15},$$
$$\sigma(3) = 5,$$
$$\sigma(n) = 1 + \frac{1}{2^n} \text{ for all } n \in N_1, n \geq 4.$$

Obviously, 10^{23} is a bound for the sequence σ since $| \sigma(n) | \leq 10^{23}$ for all $n \in N_1$. But one has the feeling that 10^{23} is not a particularly useful sort of bound since, except for the first three terms, $\frac{9}{8}$, say, is a much more modest bound for all the remaining terms of the sequence. And this smaller bound seems to have more to do with the character of the sequence than does 10^{23}. This leads to the concept of "essential bound."

* Definition 9. Let σ be a sequence. A positive rational number r is an *essential bound* for σ if and only if there exists a natural number m such that for all $n \geq m$, $| \sigma(n) | \leq r$.*

In the example preceding Definition 9, $\frac{9}{8}$ is an essential bound for the given sequence. Of course, a sequence may have many essential bounds. In the above example, $\frac{17}{16}$, $\frac{33}{32}$, etc., are essential bounds.

Clearly, a sequence may have an essential bound which is not a bound. On the other hand, every bound is an essential bound (why?).

Although a sequence may possess an essential bound which is not a bound, we can prove that every sequence possessing an essential bound is bounded. First we require a definition and a lemma.

* Definition 10. If a and b are rational numbers then *max* $\{a,b\}$ (read, "maximum of a and b") is a or b according as $a \geq b$ or $b \geq a$. (If $a \geq b$ and $b \geq a$ then $a = b$, and max $\{a,b\} = a = b$.)

lemma: If A is a finite set of rational numbers then there exists a positive rational number r such that for all $a \in A$, $| a | \leq r$.

proof: By induction. We give only a sketch of the proof since the detailed induction proof might obscure the main idea.

If A has 0 elements then $r = 1$ will do as a bound (we could also take $r = \frac{1}{2}$ or $r = 1{,}000{,}000$). Now, assume the result is true for all sets with n elements and let B be a set with $n + 1$ elements. Let $b \in B$; then $B - \{b\}$ has n elements; therefore there is a positive rational number r_1 such that for all $a \in B - \{b\}, |a| \leq r_1$. Finally, let $r = \max \{r_1, |b|\}$. Then for all $a \in B, |a| \leq r$.

q.e.d._∗

EXERCISE

Write out a detailed proof of the lemma.

Theorem 5.   If a sequence σ has an essential bound, then it is bounded.

proof: Let r be an essential bound for σ and let m be a natural number such that $|\sigma(n)| \leq r$ for all $n \geq m$. Let A be the finite set of rational numbers $\{\sigma(1),\sigma(2),\ldots,\sigma(m-1)\}$. ($A$ may contain fewer than $m - 1$ elements—namely, in case $\sigma(i) = \sigma(j)$ for some natural numbers $i, j, i \neq j, 1 \leq i \leq m - 1, 1 \leq j \leq m - 1$—but that does not matter.) Then by the lemma there is a positive rational number r' such that $|\sigma(n)| \leq r'$ if $n < m$. Then the positive rational number $\max \{r,r'\}$ is a bound for σ.

q.e.d._∗

4.3.   CANTOR SEQUENCES

We have seen that sequences of successively more precise physical measurements of an object appear to have two outstanding properties. One is the property of boundedness studied in 4.2, the other is that "later" terms of the sequence are "quite close" together. In order to study the latter property we consider a sequence which appears to have it, and see what convinces us that it does. The sequence σ defined by

(4.1)
$$\sigma(n) = 1 - \frac{1}{2^n}, \; n \in N_1$$

evidently has the property of "closeness" for "later" terms. First of all, how shall we judge "closeness"? One way is to take differences of terms; in other words, two terms are regarded as "close" if the difference between them is "small." This idea, without further qualification, could land us in a pickle. For instance, we might regard $1/10^5$ as "small". But certainly $-10^{10} < 1/10^5$. Is -10^{10} "smaller" than $1/10^5$? If two terms of a sequence differed by as little as $1/10^5$ there would be some justification for regarding them as "close." But if they differed by -10^{10}, it might be silly to consider them as close. The difficulty is eliminated easily by regarding terms as "close" if the absolute value of the difference is "small"; thus, $\sigma(n)$ and $\sigma(m)$ are close if

$$| \sigma(n) - \sigma(m) |$$

is "small."

Just how close are the terms of the sequence (4.1)? Are they within $\frac{1}{10}$ of each other? That is, is the absolute value of the difference of any two terms less than $\frac{1}{10}$? It is easy to check, for this sequence, that $| \sigma(n) - \sigma(m) | < \frac{1}{10}$ if $\sigma(n)$ and $\sigma(m)$ are any terms from the fourth on; i.e., if $n \geq 4$ and $m \geq 4$. Thus, if "closeness" is defined by $\frac{1}{10}$ and "later" is defined by "all terms from the fourth, on," then all the "later" terms of the sequence (4.1) will be "close" to each other. But one may object that $\frac{1}{10}$ isn't close enough. Will the "later" terms be within $\frac{1}{100}$ of each other? Again, it is easy to verify that if by "later" we mean "all terms from the seventh, on," then all terms will be "close" to each other with a closeness of $\frac{1}{100}$. If $\frac{1}{100}$ is not close enough, take $\frac{1}{1000}$. All terms from the tenth, on, will be within $\frac{1}{1000}$ of each other. In short,

for a "closeness" of $\frac{1}{10}$, "later" means "all terms from the 4th, on";
for a "closeness" of $\frac{1}{100}$, "later" means "all terms from the 7th, on";
for a "closeness" of $\frac{1}{1000}$, "later" means "all terms from the 10th, on".

Of course, powers of ten are not essential to the underlying idea—they are merely an accident of the fact that we possess ten fingers:

for a "closeness" of $\frac{1}{11}$, "later" means "all terms from the 4th, on";
for a "closeness" of $\frac{2}{193}$, "later" means "all terms from the 10th, on",

etc. From this analysis it appears that "closeness" and "later" are related by means of a pairing

closeness—later

where "closeness" is given by positive rational numbers and "later" by natural numbers. In turn, this idea suggests that "closeness" and "later" are combined to yield a mapping of the positive rationals into the natural numbers. We now put these ideas into a precise formulation:

*Let R^+ be the set of all positive rational numbers; thus

$$R^+ = \{\alpha \mid \alpha \in R \text{ and } \alpha > 0\}.$$

Definition 11. Let σ be a sequence of rational numbers. A mapping $\varphi : R^+ \longrightarrow N_1$ is a *convergence test* for σ if and only if for each $\alpha \in R^+$, if m, n are natural numbers such that $m \geq \varphi(\alpha)$ and $n \geq \varphi(\alpha)$ then $\mid \sigma(m) - \sigma(n) \mid < \alpha$.

A sequence which possesses a convergence test is a *Cantor sequence.*＊

The Cantor sequences are the sequences in which "later" terms are "close" together; they are the sequences in which we are primarily interested.

EXAMPLES

1. Since the discussion preceding Definition 11 was a convincing argument for the "closeness" of the "later" terms of the sequence (4.1), the formal definition, if it is any good, should also reveal this. Define the convergence test function as follows:

    for each $\alpha \in R^+$, $\alpha \geq 1$, $\varphi(\alpha) = 1$; if $\alpha < 1$, then there exists a natural number k such that $1/2^{k+1} < \alpha \leq 1/2^k$.[3] In this case, set $\varphi(\alpha) = 3k + 1$.

    Now, if $\alpha \geq 1$ then for all $n \geq 1$ and for all $m \geq 1$ (where $n \geq m$) we have

    $$\mid \sigma(m) - \sigma(n) \mid = \left| \left(1 - \frac{1}{2^m}\right) - \left(1 - \frac{1}{2^n}\right) \right|$$

    $$= \left| \frac{2^{n-m} - 1}{2^n} \right| \leq \frac{2^{n-m}}{2^n} = \frac{1}{2^m} < 1 \leq \alpha.$$

    Here, the measure of "closeness" is $\alpha \geq 1$, and "later" means "all terms from the first, on."

    Next, let $\frac{1}{2} < \alpha \leq 1$; since $k = 0$, $\varphi(\alpha) = 1$ and for all $n, m \geq 1$ (where $n \geq m$) we have

    $$\mid \sigma(m) - \sigma(n) \mid \leq \frac{1}{2^m} \leq \frac{1}{2} < \alpha.$$

[3] The proof that there is such an integer k is carried out as follows: (a) By the Archimedean Law for the rational numbers, there is a positive integer s such that $s = s \cdot 1 > 1/\alpha$. (b) By the PFI, for each integer, $n \geq 0$, there is a positive integer a such that $2^a > n$. Hence, in particular, there is a positive integer t such that $2^t > s$. Therefore, $2^t > 1/\alpha$. (c) By the WOP, there is a smallest natural number $k + 1$ such that $2^{k+1} > 1/\alpha$. Hence, $1/\alpha \geq 2^k$ and $1/2^{k+1} < \alpha \leq 1/2^k$.

Similarly, if $1/2^2 \le \alpha \le \frac{1}{2}$ then $k = 1$, $\varphi(\alpha) = 4$ and for **all** $n, m \ge 4$ (where $n \ge m$),

$$| \sigma(m) - \sigma(n) | \le \frac{1}{2^m} \le \frac{1}{2^2} < \alpha.$$

The reader should find the "later" terms for $1/2^3 < \alpha \le 1/2^2$, for $1/2^4 < \alpha \le 1/2^3$, and for arbitrary $\alpha \in R^+$, $\alpha < 1$.

2. The sequence

$$\sigma(n) = \begin{cases} 1, & n \text{ odd,} \\ \dfrac{1}{2^n}, & n \text{ even} \end{cases}$$

does not appear to have the "closeness" property for the "later" terms. For instance, if we take $\alpha = 1/10$, n odd and m even, then

$$| \sigma(m) - \sigma(n) | = \left| 1 - \frac{1}{2^m} \right| > \frac{1}{10},$$

no matter how large n and m are. Therefore it should be possible to prove that there is no convergence test for this sequence. Indeed, suppose $\varphi : R^+ \longrightarrow N_1$ is a convergence test. Then for all $\alpha \in R^+$ and for all $n \ge \varphi(\alpha)$, $m \ge \varphi(\alpha)$ it must be true that $| \sigma(m) - \sigma(n) | < \alpha$. But, as indicated above, if we take $\alpha = \frac{1}{10}$, m an even natural number $\ge \varphi(\alpha)$, n an odd natural number $\ge \varphi(\alpha)$, it follows that $| \sigma(m) - \sigma(n) | = | 1 - 1/2^n | > \frac{1}{10} = \alpha$. This is a contradiction; therefore there exists no convergence test for σ, and σ is not a Cantor sequence.

EXERCISES

1. Find another (simpler) convergence test for the sequence of Example 1. In each of Exercises 2–6 determine whether or not the given sequence is a Cantor sequence. If it is, find at least two convergence tests.

2. $\sigma(n) = 1 + 1/2^n$, $n \in N_1$.

3. $\sigma(n) = \begin{cases} 1, & \text{if } n \text{ is odd,} \\ 1 + 1/2^n, & \text{if } n \text{ is even.} \end{cases}$

4. $\sigma(n) = 1$ for all $n \in N_1$.

5. $\sigma(n) = \begin{cases} 1 - 1/2^n, & \text{if } n \text{ is odd,} \\ 1 + 1/2^n, & \text{if } n \text{ is even.} \end{cases}$

6. $\sigma(n) = 2^n, n \in N_1$.

7. Prove: If a sequence σ has a convergence test then it has infinitely many.   Hence, a convergence test φ is not uniquely determined by the sequence σ.

Definition 11 deserves further comment.   In the first place, since a sequence is a mapping $\sigma : N_1 \longrightarrow R$, the numbers of a sequence may be negative.   Since only positive rational numbers are used for purposes of measurement it follows that our concept (Definition 11) is a slight extension of the concept to which the use of measurement would have led.   However, the extension of the concept to negative rationals presents no difficulties and is mathematically quite natural and useful.

Second, if we return to sequences which arise from measurements, a little thought shows that Cantor sequences are the sort of sequences that would arise from a series of more and more precise measurements. Suppose $\tau : N_1 \longrightarrow R$ is the sequence that arises from a series of measurements.   Then we would expect that all the measurements after the fifth, say, would be within $\frac{1}{10}$ (foot) of each other, and all the measurements after the 27th, say, would be within $\frac{1}{100}$ of each other, and that all measurements after the 218th, say, would be within $\frac{3}{100,683}$ of each other, etc.   Now all the above could be summed up by saying that the sequence τ has a convergence test

$\varphi : R^+ \longrightarrow N_1$ such that

$$\varphi\left(\frac{1}{10}\right) = 5, \quad \varphi\left(\frac{1}{100}\right) = 27, \quad \varphi\left(\frac{3}{100,683}\right) = 218.$$

Finally, notice the presence of the absolute value signs in Definition 11.   Remember that the statement $| \varphi(m) - \varphi(n) | < \alpha$ means $-\alpha < \varphi(m) - \varphi(n) < \alpha$.   This is as it should be.   We do not care whether the successive numbers in the sequence increase, or decrease, or fluctuate; what we do care about is that they are eventually all close together.

Since the properties of boundedness and of being a Cantor sequence occur for sequences arising from physical measurements, it is reasonable to inquire how these properties are related.   Example 2, page 190, shows that a bounded sequence need not be a Cantor sequence.   On the other hand, we can prove

* Theorem 6. Every Cantor sequence is bounded.

proof: Let σ be a Cantor sequence and let $\varphi : R^+ \longrightarrow N_1$ be a convergence test for σ. According to Theorem 5, if we can prove that σ has an essential bound, then σ has a bound.

We claim that $|\sigma(\varphi(\frac{5}{9}))| + \frac{5}{9}$ is an essential bound for σ. For, let $m = \varphi(\frac{5}{9})$ and let n be any natural number $\geq m$. Then

$$| \sigma(n) | = | \sigma(m) + (\sigma(n) - \sigma(m)) |$$
$$\leq | \sigma(m) | + | \sigma(n) - \sigma(m) |.$$

But $| \sigma(m) | = | \sigma(\varphi(\frac{5}{9})) |$ since $m = \varphi(\frac{5}{9})$. Also, since $n, m \geq \varphi(\frac{5}{9})$, we have $| \sigma(n) - \sigma(m) | < \frac{5}{9}$. Hence,

$$| \sigma(m) | + | \sigma(n) - \sigma(m) | < | \sigma(\varphi(\frac{5}{9})) | + \frac{5}{9};$$

therefore,

$$| \sigma(n) | \leq | \sigma(\varphi(\frac{5}{9})) | + \frac{5}{9}$$

for all $n \geq m = \varphi(\frac{5}{9})$. By Definition 9, $| \sigma(\varphi(\frac{5}{9})) | + \frac{5}{9}$ is an essential bound, hence (Theorem 5) σ is bounded.

q.e.d.*

Of course, the choice of the positive rational number $\frac{5}{9}$ in the proof of Theorem 6 was purely arbitrary. For instance, 10,682,941,735 would have done as well. Most authors would use 1 since a "1" is easiest to write.

If we let $\mathcal{C}$ be the set of Cantor sequences then Theorem 6 can be restated as

* Theorem 6'. $\mathcal{C} \subset \mathcal{B}.$

Theorem 7. Every constant sequence is a Cantor sequence.

proof: Let $\hat{r}$ be a constant sequence and let $\varphi : R^+ \longrightarrow N_1$ be any mapping whatever. Then, φ is a convergence test for $\hat{r}$. Indeed, if $\alpha \in R^+$ and $n, m \geq \varphi(\alpha)$ then

$$| \hat{r}(n) - \hat{r}(m) | = | r - r | = | 0 | = 0 < \alpha.$$

q.e.d.*

The next theorem tells us something about the arithmetic of Cantor sequences.

Theorem 8. Sums, differences and products of Cantor sequences are Cantor sequences. Moreover, properties 1–8 of Theorem 3 hold for Cantor sequences. (Hence, $\mathcal{C}$ is a commutative ring with unity element. In fact, $\mathcal{C}$ is a subring of both $\mathcal{S}$ and $\mathcal{B}$.)

proof: Let γ_1, $\gamma_2 \in \mathcal{C}$; we prove that $\gamma_1 + \gamma_2 \in \mathcal{C}$. Let φ_1, φ_2 be convergence tests for γ_1, γ_2, respectively. Define $\varphi : R^+ \longrightarrow N_1$ by

$$\varphi(r) = \max \left\{ \varphi_1 \left(\frac{r}{2} \right), \varphi_2 \left(\frac{r}{2} \right) \right\}, \quad r \in R^+;$$

φ is a mapping of R^+ into N_1 (why?). We assert that φ is a convergence test for $\gamma_1 + \gamma_2$, hence $\gamma_1 + \gamma_2$ is a Cantor sequence. Therefore we must prove that for each $\alpha \in R^+$, if n, $m \geq \varphi(\alpha)$ then

$$| (\gamma_1 + \gamma_2)(n) - (\gamma_1 + \gamma_2)(m) | < \alpha.$$

If n, $m \geq \varphi(\alpha)$ then (by definition of φ) n, $m \geq \varphi_1(\alpha/2)$ and n, $m \geq \varphi_2(\alpha/2)$. Since φ_1, φ_2 are convergence tests for γ_1, γ_2, respectively, it follows that

$$| \gamma_1(n) - \gamma_1(m) | < \frac{\alpha}{2} \text{ and } | \gamma_2(n) - \gamma_2(m) | < \frac{\alpha}{2}.$$

But now

$$\begin{aligned}
| (\gamma_1 + \gamma_2)(n) - (\gamma_1 + \gamma_2)(m) | &= | (\gamma_1(n) + \gamma_2(n)) - (\gamma_1(m) + \gamma_2(m)) | \\
&= | (\gamma_1(n) - \gamma_1(m)) + (\gamma_2(n) - \gamma_2(m)) | \\
&\leq | \gamma_1(n) - \gamma_1(m) | + | \gamma_2(n) - \gamma_2(m) | \\
&< \frac{\alpha}{2} + \frac{\alpha}{2} = \alpha,
\end{aligned}$$

for each $\alpha \in R^+$. Therefore, φ is a convergence test for $\gamma_1 + \gamma_2$.

We leave it to the reader to prove that if $\gamma \in \mathcal{C}$ then $-\gamma \in \mathcal{C}$. With this result and the foregoing it follows at once that $\gamma_1 - \gamma_2 = \gamma_1 + (-\gamma_2) \in \mathcal{C}$.

In order to prove that $\gamma_1 \cdot \gamma_2 \in \mathcal{C}$ we make use of the fact that $\mathcal{C} \subset \mathcal{F}$, i.e., that every Cantor sequence is bounded. As before, let φ_1, φ_2 be convergence tests for γ_1, γ_2 and further let b_1, b_2 be bounds for γ_1 and γ_2, respectively. This time define $\varphi : R^+ \longrightarrow N_1$ by

$$\varphi(r) = \max \left\{ \varphi_1 \left(\frac{r}{2b_2} \right), \varphi_2 \left(\frac{r}{2b_1} \right) \right\};$$

again it is easy to see that φ is a mapping. We propose to show that φ is a convergence test for $\gamma_1 \cdot \gamma_2$. Suppose n, $m \geq \varphi(\alpha)$, $\alpha \in R^+$; then n, $m \geq \varphi_1(\alpha/2b_2)$ and n, $m \geq \varphi_2(\alpha/2b_1)$. Hence,

$$| (\gamma_1 \cdot \gamma_2)(n) - (\gamma_1 \cdot \gamma_2)(m) |$$
$$= | \gamma_1(n) \cdot \gamma_2(n) - \gamma_1(m) \cdot \gamma_2(m) |$$
$$= | \gamma_1(n)(\gamma_2(n) - \gamma_2(m)) + \gamma_2(m)(\gamma_1(n) - \gamma_1(m)) |$$
$$\leq | \gamma_1(n) | \cdot | \gamma_2(n) - \gamma_2(m) | + | \gamma_2(m) | \cdot | \gamma_1(n) - \gamma_1(m) |$$
$$< b_1 \cdot \frac{\alpha}{2b_1} + b_2 \cdot \frac{\alpha}{2b_2} = \frac{\alpha}{2} + \frac{\alpha}{2} = \alpha.$$

Finally, since $\hat{0}$ and $\hat{1} \in \mathcal{C}$, and since properties 1–8 of Theorem 3 hold for all sequences, in particular these properties must hold for the sequences of $\mathcal{C}$.

q.e.d.*

EXERCISE

Let σ be a sequence such that

(i)   there exists a natural number n_0 and a positive rational number K such that for all $n \geq r_0$, $\sigma(n) \geq K$ and

(ii) $\sigma^2 = \sigma \cdot \sigma$ is a Cantor sequence.
      Prove that σ is a Cantor sequence.

4.4.   NULL SEQUENCES

Let us return to a problem stated at the beginning of this chapter. A right isosceles triangle is constructed and it is proposed to measure the length of the hypotenuse with ever-increasing precision. The task is assigned to a person who, we assume, makes no mistakes in observation or calculation and who has a set of measuring instruments appropriate to the task. As a result of his measurements he obtains an infinite sequence σ of rational numbers whose first six terms are

$$\frac{3}{2}, \quad \frac{11}{8}, \quad \frac{23}{16}, \quad \frac{179}{128}, \quad \frac{365}{250}, \quad \frac{1439}{1024}.$$

(Of course we are indulging in fantasy here, but this is useful in clarifying the idea we have in mind.) Since the only numbers available thus far are the natural numbers, the integers and the rationals, and since $\sqrt{2}$ is not rational (hence is not an integer or a natural number), we might decide to define $\sqrt{2}$ as the sequence σ. Next, suppose that a

second person, satisfying all the requirements of the first, is assigned to the same job. Even under the most favorable circumstances the sequences obtained by the two persons will not be the same. For instance, the second person might obtain a sequence τ whose first six terms are

$$\frac{3}{2}, \quad \frac{23}{16}, \quad \frac{91}{64}, \quad \frac{1451}{1024}, \quad \frac{5797}{4096}, \quad \frac{46,355}{32,768}.$$

Now we are faced with a quandary. For, under our hypotheses, there is no reason to define $\sqrt{2}$ as the sequence σ in preference to the sequence τ. We can get out of our difficulty by deciding that $\sqrt{2}$ will be defined, not by a single sequence, but rather by all sequences "like" σ and τ. This results in shifting the problem to that of determining a useful concept of "like" sequences. As a hint, notice that $\sigma - \tau$ is a sequence whose first six terms are

$$0, \quad \frac{-1}{10}, \quad \frac{1}{64}, \quad \frac{-19}{1024}, \quad \frac{43}{4096}, \quad \frac{-307}{32,768},$$

and apparently the "later" terms of $\sigma - \tau$ are closer and closer to zero. This leads to the suggestion that two sequences σ and τ are "like" one another if the "later" terms of $\sigma - \tau$ are closer and closer to zero. Consequently, the problem of characterizing "like" sequences is reduced to that of characterizing sequences with the property that the "later" terms are closer and closer to zero. These will be called "null" sequences.

* Definition 12. A sequence σ is a *null* sequence if and only if there exists a mapping $\varphi : R^+ \longrightarrow N_1$ such that for each $a \in R^+$ if $n \geq \varphi(a)$ then $|\sigma(n)| \leq a$. A mapping φ satisfying these conditions is a *nullity test for* σ. The set of all null sequences is denoted by "$\mathfrak{N}$".*

The definition of null sequence resembles so closely that of Cantor sequence that one is led to conjecture

* Theorem 9. Every null sequence is a Cantor sequence; i.e., $\mathfrak{N} \subset \mathfrak{C}$.

proof: Let $\sigma \in \mathfrak{N}$ and let $\varphi : R^+ \longrightarrow N_1$ be a nullity test for σ. The theorem is proved by exhibiting a convergence test, $\psi : R^+ \longrightarrow N_1$ for σ. Define ψ by

$$\psi(\alpha) = \varphi\left(\frac{\alpha}{2}\right), \text{ for all } \alpha \in R^+.$$

We assert that ψ is a convergence test for σ. For all $n, m \in N_1$ we have

$$| \sigma(n) - \sigma(m) | = | \sigma(n) + (-\sigma(m)) |$$
$$\leq | \sigma(n) | + | -\sigma(m) |$$
$$= | \sigma(n) | + | \sigma(m) |.$$

Now let $n, m \geq \psi(\alpha)$; then $n, m \geq \varphi(\alpha/2)$. But since φ is a nullity test for σ and since $n, m \geq \varphi(\alpha/2)$, it follows that $| \sigma(n) | < \alpha/2$ and $| \sigma(m) | < \alpha/2$. Hence, for all $n, m \geq \psi(\alpha)$,

$$| \sigma(n) - \sigma(m) | < \frac{\alpha}{2} + \frac{\alpha}{2} = \alpha.$$

Since this inequality holds for each $\alpha \in R^+$ the theorem is proved.

q.e.d.*

EXERCISES

1. Which constant sequences, if any, are null sequences?

2. Define several null sequences and find nullity tests for them.

3. Prove: If a sequence has one nullity test, it has infinitely many.

4. Prove: If $\nu \in \mathfrak{N}$ then $-\nu \in \mathfrak{N}$.

One expects that the arithmetic of $\mathfrak{N}$ will be similar to the arithmetic of $\mathfrak{C}$, $\mathfrak{B}$ and $\mathfrak{S}$. It is; but $\mathfrak{N}$ exhibits one new, important property when considered as a subset of $\mathfrak{B}$ and of $\mathfrak{C}$.

* **Theorem 10.** (a) Sums and differences of null sequences are null sequences.

(b) for all $\nu \in \mathfrak{N}$ and for all $\beta \in \mathfrak{B}$, $\beta \cdot \nu = \nu \cdot \beta \in \mathfrak{N}$.

(b') For all $\nu \in \mathfrak{N}$ and for all $\gamma \in \mathfrak{C}$, $\gamma \cdot \nu = \nu \cdot \gamma \in \mathfrak{N}$.

((b) and (b') are the properties of $\mathfrak{N}$ relative to $\mathfrak{B}$ and $\mathfrak{C}$, respectively, mentioned above.)

(c) Properties 1–8 of Theorem 3 hold for $\mathfrak{N}$. (However, since $\hat{1} \notin \mathfrak{N}$, $\mathfrak{N}$ is a commutative ring but *not* with unity element).

proof: (a) Let $\nu_1, \nu_2 \in \mathfrak{N}$ and let φ_1, φ_2 be nullity tests for ν_1, ν_2 respectively. Define $\varphi : R^+ \longrightarrow N_1$ by

$$\varphi(\alpha) = \max\left(\varphi_1\left(\frac{\alpha}{2}\right), \varphi_2\left(\frac{\alpha}{2}\right) \right),$$

for all $\alpha \in R^+$. Then if $\alpha \in R^+$ and n is a natural number, $n \geq \varphi(\alpha)$, we have $n \geq \varphi_1(\alpha/2)$, $n \geq \varphi_2(\alpha/2)$, whence

$$| (\nu_1 + \nu_2)(n) | = | \nu_1(n) + \nu_2(n) | \leq | \nu_1(n) | + | \nu_2(n) | < \frac{\alpha}{2} + \frac{\alpha}{2} = \alpha;$$

therefore φ is a nullity test for $\nu_1 + \nu_2$. Similarly, the same φ is a nullity test for $\nu_1 - \nu_2$.

(b) Since $\nu \in \mathfrak{N}$ it has a nullity test φ. And since $\beta \in \mathfrak{B}$ it has a bound $c \neq 0$. To prove that $\nu \cdot \beta \in \mathfrak{N}$ we exhibit a nullity test for this sequence. Define $\psi: R^+ \longrightarrow N_1$ by

$$\psi(\alpha) = \varphi\left(\frac{\alpha}{c}\right)$$

for all $\alpha \in R^+$. We claim that ψ is a nullity test for $\nu \cdot \beta$. For, if n is a natural number, $n \geq \psi(\alpha)$, then $n \geq \varphi(\alpha/c)$ so that

$$| (\nu \cdot \beta)(n) | = | \nu(n) \cdot \beta(n) | = | \nu(n) | \cdot | \beta(n) | < \frac{\alpha}{c} \cdot c = \alpha,$$

and this proves our assertion. Since a product of sequences is commutative, ψ is also a nullity test for $\beta \cdot \nu$.

(b') Since $\mathfrak{C} \subset \mathfrak{F}$, (b') follows at once.

(c) This duplicates reasoning presented elsewhere and is left to the reader.

q.e.d.*

Readers familiar with the terminology of abstract algebra will recognize that (b) and (b') in Theorem 10, together with (a), characterize $\mathfrak{N}$ as an ideal in $\mathfrak{B}$ and $\mathfrak{C}$, respectively.

EXERCISES

1. Is it true that for all $\gamma \in \mathfrak{C}$ and for all $\beta \in \mathfrak{B}$, $\gamma \cdot \beta \in \mathfrak{C}$? If not, give a counter-example.

2. Do the same as Exercise 1 for $\mathfrak{C}$ and $\mathfrak{S}$ and for $\mathfrak{B}$ and $\mathfrak{S}$.

3. Compute the product of β and ν where $\beta(n) = \begin{cases} 1, & n \text{ odd}, \\ -1, & n \text{ even} \end{cases}$ and $\nu(n) = 1/2^n$, $n \in N_1$. Find a nullity test for $\nu \cdot \beta$.

4. For the same sequence ν as in Exercise 3, and for the sequence $\gamma(n) = 1 + 1/2^n$, $n \in N_1$, find a nullity test for $\nu \cdot \gamma$.

The concept of "like" Cantor sequences is made exact in

* Definition 13. For all Cantor sequences γ_1, γ_2, $\gamma_1 \sim \gamma_2$ if and only if $\gamma_1 - \gamma_2$ is a null sequence.*

Definition 13 constitutes a decisive step in the construction of the real numbers.

EXERCISE

State Definition 13 in terms of a subset of $\mathcal{C} \times \mathcal{C}$.

* Theorem 11. The relation $\sim$ is an equivalence relation on $\mathcal{C}$.

proof: (1) Let γ be any sequence in $\mathcal{C}$, then $\gamma - \gamma = \hat{0}$ and since $\hat{0} \in \mathfrak{N}$ it follows that $\gamma \sim \gamma$ for all $\gamma \in \mathcal{C}$. Therefore, $\sim$ is reflexive.

(2) Suppose γ_1, $\gamma_2 \in \mathcal{C}$ and $\gamma_1 \sim \gamma_2$. Then $\gamma_1 - \gamma_2$ is a null sequence, therefore (Theorem 10 (a)), so is $-(\gamma_1 - \gamma_2) = \gamma_2 - \gamma_1$. Hence, $\gamma_2 \sim \gamma_1$, and $\sim$ is symmetric.

(3) Finally, let γ_1, γ_2, $\gamma_3 \in \mathcal{C}$ where $\gamma_1 \sim \gamma_2$ and $\gamma_2 \sim \gamma_3$. Then $\gamma_1 - \gamma_2$ and $\gamma_2 - \gamma_3$ are null sequences, hence (Theorem 10 (a)), so is $(\gamma_1 - \gamma_2) + (\gamma_2 - \gamma_3) = \gamma_1 - \gamma_3$. Consequently, $\gamma_1 \sim \gamma_3$, i.e., $\sim$ is transitive.

By Definition 28', Chapter 1, $\sim$ is an equivalence relation.

q.e.d.*

From Theorem 11 (above) and from Theorem 12, Chapter 1, we deduce immediately

* Theorem 12. The equivalence relation $\sim$ determines a partition of $\mathcal{C}$.*

For emphasis we state below some of the essential properties of the equivalence classes of this partition (see Section 1.15, Chapter 1):

1. each class is a nonempty set of Cantor sequences;

2. each Cantor sequence is an element of an equivalence class, hence the union of the equivalence classes is $\mathcal{C}$;

3. if $\bar{\alpha}$ is a class containing the sequence α, $\bar{\beta}$ a class containing β, then exactly one of $\bar{\alpha} = \bar{\beta}$ or $\bar{\alpha} \cap \bar{\beta} = \phi$ holds;

4. $\bar{\alpha} = \bar{\beta}$ if and only if $\alpha - \beta$ is a null sequence.

notations: The set of equivalence classes will be denoted by "$\mathbb{R}$". If $\gamma \in \mathcal{C}$ and γ is not a constant sequence then the class containing the element γ will be denoted by "$\bar{\gamma}$". Consistency would demand that since $\hat{r}$ is a constant sequence the equivalence class containing $\hat{r}$ be denoted by "$\bar{\hat{r}}$". Since this notation is clumsy we define

$$\bar{r} = \bar{\hat{r}}$$

for all rational numbers r.

Definition 14. $\mathbb{R}$ is the set of *real numbers.*

At this stage the reader may wonder why the elements of $\mathbb{R}$ are called "real numbers" and, in particular, what does the present concept have to do with his intuitive ideas of real numbers.   We tax his patience just a little longer.   After sums and products have been defined it will be possible to illustrate the foregoing theory in a way that justifies considering the elements of $\mathbb{R}$ as real numbers.

4.5.   THE REAL NUMBERS

The next step is the development of an arithmetic for the real numbers (the reader should review the definition of "binary operation," Definition 27, Chapter 1).

If $\alpha, \beta \in \mathcal{C}$ then $\alpha + \beta$ is also an element in $\mathcal{C}$, hence $\alpha + \beta$ determines an equivalence class $\overline{\alpha + \beta}$,

$$\overline{\alpha + \beta} = \{\gamma \mid \gamma \in \mathcal{C} \text{ and } \gamma \sim \alpha + \beta\}.$$

Similarly, $\overline{\alpha \cdot \beta}$ is the equivalence class

$$\overline{\alpha \cdot \beta} = \{\delta \mid \delta \in \mathcal{C} \text{ and } \delta \sim \alpha \cdot \beta\}.$$

The next theorem is essential for the definition of the binary operations of "multiplication" and "addition" of real numbers.

* **Theorem 13.**   If $\alpha, \beta, \gamma, \delta$ are Cantor sequences such that $\bar{\alpha} = \bar{\gamma}$ and $\bar{\beta} = \bar{\delta}$ then

(a) $\overline{\alpha + \beta} = \overline{\gamma + \delta}$, and
(b) $\overline{\alpha \cdot \beta} = \overline{\gamma \cdot \delta}$.

proof: From $\bar{\alpha} = \bar{\gamma}$ and $\bar{\beta} = \bar{\delta}$ it follows that $\alpha - \gamma$ and $\beta - \delta \in \mathfrak{N}$.
(a) By Theorem 10(a), $(\alpha - \gamma) + (\beta - \delta) \in \mathfrak{N}$, and by the commu-

tativity of $+$ it follows that $(\alpha + \beta) - (\gamma + \delta) \in \mathfrak{N}$. Hence, $\overline{\alpha + \beta} = \overline{\gamma + \delta}$.

$$(b) \quad (\alpha \cdot \beta) - (\gamma \cdot \delta) = ((\alpha \cdot \beta) - (\alpha \cdot \delta)) + ((\alpha \cdot \delta) - (\gamma \cdot \delta))$$
$$= \alpha \cdot (\beta - \delta) + (\alpha - \gamma) \cdot \delta.$$

Since $\beta - \delta \in \mathfrak{N}$ and $\alpha \in \mathcal{C}$, we deduce from Theorem 10(b) that $\alpha \cdot (\beta - \delta) \in \mathfrak{N}$. Similarly, $\overline{(\alpha - \gamma) \cdot \delta} \in \mathfrak{N}$. Therefore, $(\alpha \cdot \beta) - (\gamma \cdot \delta) \in \mathfrak{N}$, whence $\overline{\alpha \cdot \beta} = \overline{\gamma \cdot \delta}$.

q.e.d.

Definition 15. $\oplus$ is the subset of $(\mathbb{R} \times \mathbb{R}) \times \mathbb{R}$ consisting of all $((\bar\alpha, \bar\beta), \bar\gamma)$ such that $\bar\gamma = \overline{\alpha + \beta}$. $\odot$ is the subset of $(\mathbb{R} \times \mathbb{R}) \times \mathbb{R}$ consisting of all $((\bar\alpha, \bar\beta), \bar\delta)$ such that $\bar\delta = \overline{\alpha \cdot \beta}$.

Theorem 14. $\oplus$ and $\odot$ are binary operations on $\mathbb{R}$.

proof: We prove that $\oplus$ and $\odot$ are mappings of $\mathbb{R} \times \mathbb{R}$ into $\mathbb{R}$.

To show that $\oplus$ is a mapping it suffices to prove: if $((\bar\alpha, \bar\beta), \bar\gamma)$ and $((\bar\alpha_1, \bar\beta_1), \bar\gamma_1)$ are elements in $\oplus$ and if $(\bar\alpha, \bar\beta) = (\bar\alpha_1, \bar\beta_1)$ then $\bar\gamma = \bar\gamma_1$. But $(\bar\alpha, \bar\beta) = (\bar\alpha_1, \bar\beta_1)$ means that $\bar\alpha = \bar\alpha_1$ and $\bar\beta = \bar\beta_1$, whence (Theorem 13(a)) $\bar\gamma = \overline{\alpha + \beta} = \overline{\alpha_1 + \beta_1} = \bar\gamma_1$.

The proof that $\odot$ is a binary operation is similar to the above and is left as an exercise.

q.e.d.

Definition 16. The binary operation $\oplus$ is the *sum* of real numbers; $\odot$ is the *product* of real numbers. If $((\bar\alpha, \bar\beta), \bar\gamma) \in \oplus$ we write "$\bar\gamma = \bar\alpha \oplus \bar\beta$"; if $((\bar\alpha, \bar\beta), \bar\delta) \in \odot$ we write "$\bar\delta = \bar\alpha \odot \bar\beta$".

corollary: $\bar\alpha \oplus \bar\beta = \overline{\alpha + \beta}$ and $\bar\alpha \odot \bar\beta = \overline{\alpha \cdot \beta}$ for all α, $\beta \in \mathcal{C}$.

proof: Definitions 15 and 16. ∗

EXAMPLES

1. The equivalence class (or, real number) $\bar{2}$ containing the constant sequence $\hat{2}$ is usually called "the real number, two." Obviously, $\bar{2}$

contains more than the constant sequence $\hat{2}$; for instance, the sequences $\sigma(n) = 2 - 1/2^n$, $n \in N_1$, and $\tau(n) = 2 + 1/2^n$, $n \in N_1$, are elements of $\bar{2}$. Later, when we prove that there is an isomorphism between R and a certain subset of R, it will be seen that under the particular isomorphism the (rational) integer two and the real number two correspond to each other. This justifies using "two" as a name for the real number $\bar{2}$.

Similarly, the equivalence class $\bar{3}$ containing the constant sequence $\hat{3}$ is called "the real number, three." According to the corollary following Definition 16, the product of $\bar{2}$ and $\bar{3}$ is $\bar{2} \odot \bar{3} = \overline{\hat{2} \cdot \hat{3}} = \overline{\hat{6}} = \bar{6}$, which conforms to our experience with real numbers—two times three is truly equal to six.

2. Since the whole fuss concerning the real numbers was started by raising questions concerning the existence of a square root of two, we now prove that the real numbers, as we have defined them, do contain an element u whose square, $u \odot u$, is $\bar{2}$.

Consider the sequence σ defined recursively by

$$\sigma(1) = 1,$$

$$\sigma(k + 1) = \frac{\sigma(k) + \dfrac{2}{\sigma(k)}}{2}, \quad k \in N_1;$$

as a matter of convenience write $x_n = \sigma(n)$ for all $n \in N_1$, so that the definition of the sequence becomes

$$x_1 = 1,$$

$$x_{k+1} = \frac{x_k + \dfrac{2}{x_k}}{2} = \frac{x_k^2 + 2}{2x_k}, \quad k \in N_1.$$

We shall prove that σ is a Cantor sequence having the property that

$$\sigma^2 = \sigma \cdot \sigma = \hat{2}$$

and therefore,

$$\bar{\sigma} \odot \bar{\sigma} = \bar{2}.$$

Thus, $\bar{\sigma}$ may be called, properly, a "square root of two." The proof will be carried out in several steps, a few being left to the reader.

(a) First, note that for all $k \in N_1$, $x_k > 0$, and further, $x_k^2 \neq 2$, since each $x_k \in R$. We leave the details to the reader.

(b) For all natural numbers $k \geq 1$, $x_{k+1}^2 - 2 = \left(\dfrac{x_k^2 - 2}{2x_k}\right)^2$.

proof:

$$x_{k+1}^2 - 2 = \left(\frac{x_k^2 + 2}{2x_k}\right)^2 - 2$$

$$= \left(\frac{x_k^2 - 2}{2x_k}\right)^2.$$

(c) For all natural numbers $k \geq 2$, $x_k^2 > 2$.

proof: This is an immediate consequence of (a) and (b).

(d) Each $x_k \geq 1$.

proof: Exercise.

(e) For all natural numbers $k \geq 1$,

$$0 < x_{k+1}^2 - 2 \leq \left(\frac{1}{4}\right)^{2^{k-1}}.$$

proof: That $x_{k+1}^2 - 2 > 0$ is the result (c). The second inequality is proved by induction.

If $k = 1$, then $x_2^2 - 2 = \dfrac{9}{4} - 2 = \dfrac{1}{4} \leq \left(\dfrac{1}{4}\right)^{2^{1-1}}$.

Let $n \geq $ | and assume

$$x_n^2 - 2 \leq \left(\frac{1}{4}\right)^{2^{n-1}-1}.$$

Then, by (b),

$$x_{n+1}^2 - 2 = \left(\frac{x_n^2 - 2}{2x_n}\right)^2 = \frac{(x_n^2 - 2)^2}{4x_n^2} \leq \frac{(x_n^2 - 2)^2}{4},$$

since (by (d)) $x_n \geq 1$. Then the induction hypothesis yields

$$x_{n+1}^2 - 2 \leq \frac{1}{4} \cdot (x_n^2 - 2)^2 \leq \frac{1}{4} \cdot \left(\left(\frac{1}{4}\right)^{2^{n-1}-1}\right)^2 = \left(\frac{1}{4}\right)^{2^n - 1}.$$

Hence, $x_{k+1}^2 - 2 \leq \left(\dfrac{1}{4}\right)^{2^{k-1}}$ holds for all natural numbers $k \geq 1$.

(f) If $\alpha \in R^+$, $\alpha < 1$, there is one and only one natural number k such that

$$\left(\frac{1}{4}\right)^{2^k - 1} < \alpha \leq \left(\frac{1}{4}\right)^{2^{k-1}-1}.$$

proof: Exercise.   (Hint: show that there is a smallest natural number k such that $4^{2^{k}-1} \geq 1/\alpha$.   Also, see footnote 3, page 189).

(g)   The sequence $\sigma^2 - \hat{2}$ is a null sequence.

proof: We construct a nullity test, $\eta : R^+ \longrightarrow N_1$, for $\sigma^2 - \hat{2}$ as follows:

if $\alpha > 1$ set $\eta(\alpha) = 1$;
if $\alpha \leq 1$ then by (f) there is a natural number k such that

$\left(\dfrac{1}{4}\right)^{2^{k}-1} < \alpha$.   In this case, set $\eta(\alpha) = k + 1$.

Now, if $\alpha > 1$ then for all $n \geq \eta(\alpha) = 1$,

$$| x_n^2 - 2 | \leq \left(\frac{1}{4}\right)^{2^{n-1}-1} \leq 1 < \alpha.$$

If $\alpha \leq 1$ then for all $n \geq \eta(\alpha) = k + 1$,

$$| x_n^2 - 2 | \leq \left(\frac{1}{4}\right)^{2^{n-1}-1} \leq \left(\frac{1}{4}\right)^{2^{k}-1} < \alpha.$$

Hence η is a nullity test for $\sigma^2 - \hat{2}$.

(h)   Since $\sigma^2 - \hat{2}$ is a null sequence it follows that σ^2 is a Cantor sequence.

(i) σ is a Cantor sequence.

proof: This follows from the exercise, page 194.

The argument is now concluded as follows:

Since σ is a Cantor sequence it determines a real number $\bar{\sigma}$.   By (g),

$$\bar{\sigma} \odot \bar{\sigma} = \bar{2},$$

i.e., the real number $\bar{\sigma}$ is a square root of two.
It is easy to verify that $\overline{-\sigma}$ is also a square root of two.

EXERCISES

1.   Using the sequence $\sigma(1) = 1$, $\sigma(n + 1) = \dfrac{\sigma(n) + \dfrac{3}{\sigma(n)}}{2}$,

$n \in N_1$, prove that R contains a square root of three.

2. Let α be a positive rational number. Use the sequence $\mu(1) = 1$,

$$\mu(n + 1) = \frac{\mu(n) + \dfrac{\alpha}{\mu(n)}}{2}, \quad n \in N_1,$$

to prove that R contains a square root of α. (Note: the only cases of interest are those in which α is not a square of a rational number.)

3. Use the sequence τ defined by

$$\tau(1) = 1, \quad \tau(n + 1) = \frac{1}{\tau(n) + 1} + 1, \quad n \in N_1,$$

to prove that R contains a "square root of two."

* Theorem 15. R has the properties 1–8 of Theorem 3 with respect to the binary operations $\oplus$ and $\odot$. (Hence, R is a commutative ring with unity element, $\bar{1}$.)

proof: 1. $\oplus$ is associative. For

$$\begin{aligned}
(\bar\alpha \oplus \bar\beta) \oplus \bar\gamma &= \overline{\alpha + \beta} \oplus \bar\gamma && \text{(Corollary to Definition 16)} \\
&= \overline{(\alpha + \beta) + \gamma} && \text{(Corollary to Definition 16)} \\
&= \overline{\alpha + (\beta + \gamma)} && \text{(Associativity of $+$)} \\
&= \bar\alpha \oplus \overline{\beta + \gamma} \\
&= \bar\alpha \oplus (\bar\beta \oplus \bar\gamma).
\end{aligned}$$

2. $\oplus$ is commutative. (Exercise.)

3. For each $\alpha \in R$, $\bar\alpha \oplus \bar{0} = \bar\alpha$. For, $\bar\alpha \oplus \bar{0} = \overline{\alpha + 0} = \bar\alpha$ since $\alpha + \hat{0} = \alpha$ (Theorem 3).

4. For each $\bar\alpha \in R$, $-\bar\alpha = \overline{(-\alpha)}$. For

$$\bar\alpha \oplus (-\bar\alpha) = \overline{\alpha + (-\alpha)} = \bar{\hat{0}} = \bar{0}.$$

We leave 5, 6 and 7 as exercises with the reminder that $\bar{1}$ is the unity element of R.

8. For all $\bar\alpha$, $\bar\beta$, $\bar\gamma$ in R,

$$\begin{aligned}
\bar\alpha \odot (\bar\beta \oplus \bar\gamma) &= \bar\alpha \odot \overline{(\beta + \gamma)} \\
&= \overline{\alpha \cdot (\beta + \gamma)} \\
&= \overline{\alpha \cdot \beta + \alpha \cdot \gamma} && \text{(Distributive law in $\mathcal{C}$.)} \\
&= \overline{\alpha \cdot \beta} \oplus \overline{\alpha \cdot \gamma} \\
&= \bar\alpha \odot \bar\beta \oplus \bar\alpha \odot \bar\gamma.
\end{aligned}$$

q.e.d.$_*$

Besides properties 1–8, it will be proved later that each real number $\bar{\alpha} \neq \bar{0}$ has a reciprocal or multiplicative inverse, $\bar{\beta} \in \mathbb{R}$, such that $\bar{\alpha} \odot \bar{\beta} = \bar{1}$, hence $\mathbb{R}$ is a field. Before obtaining this result we introduce an order relation on $\mathbb{R}$; this relation is similar to the one defined on the rationals R. We begin with

* **Definition 17.** A Cantor sequence γ is *positive* if and only if there exists a positive rational number r and a natural number m such that for all natural numbers $n \geq m$, $\gamma(n) \geq r$.*

EXAMPLES

1. $\sigma(n) = 1 - 1/2^n$, $n \in N_1$. We show that there is a positive rational number r and a natural number m such that for all $n \geq m$, $1 - 1/2^n \geq r$.

    Take $r = \dfrac{99}{100}$. Then

    $$1 - \frac{1}{2^n} \geq \frac{99}{100}$$

    if and only if

    $$2^n \geq 100.$$

    Hence, if we take $m = 7$, then for all $n \geq 7$, $2^n \geq 100$, whence $1 - 1/2^n \geq 99/100$.

2. Let the sequence σ be defined by

    $$\sigma(n) = 1 + \sum_{k=1}^{n} \frac{1}{k!}, \quad n \in N_1.$$

    If we take $r = \frac{65}{24}$ then it is easy to check that $\sigma(4) \geq \frac{65}{24}$. Since, for all n, $\sigma(n + 1) \geq \sigma(n)$ it follows that for all $n \geq 4 = m$, $\sigma(n) \geq \frac{65}{24}$.

3. Define the sequence γ by $\gamma(n) = \begin{cases} \dfrac{1}{2^n}, & n \text{ even}, \\[2mm] -\dfrac{1}{2^n}, & n \text{ odd}. \end{cases}$

    We claim that γ is not positive. Hence, it must be proved that for all rationals $r > 0$ and for every natural number m there exists an $n \geq m$ such that $\gamma(n) < r$. Obviously, whenever n is odd and $\geq$ any given m, $\gamma(n) = -1/2^n < 0 < r$.

EXERCISE

Prove that the sequence μ defined by

$$\mu(n) = 1 + \sum_{k=1}^{n} \frac{\frac{1}{2}(\frac{1}{2} - 1) \ldots (\frac{1}{2} - k)}{(k + 1)!}, \quad n \in N_1,$$

is positive. (Hint: prove that

$$\left| \frac{\frac{1}{2}(\frac{1}{2} - 1) \ldots (\frac{1}{2} - k)}{(k + 1)!} \right| \geq \left| \frac{\frac{1}{2}(\frac{1}{2} - 1) \ldots (\frac{1}{2} - (k + 1))}{(k + 2)!} \right|.$$

Note: If a sequence γ is positive, not all of the terms of γ need be $\geq r$; some at the beginning may even be negative. But "essentially" all the terms of γ are $\geq r$. For instance, the sequence η defined by $\eta(1) = -5$, $\eta(2) = -17$, $\eta(3) = -1$, $\eta(n) = 1 - 1/2^n$, $n \geq 4$ is positive (why?).

* Definition 18. Let γ be a Cantor sequence. An ordered pair $(m,r) \in N \times R^+$, such that for all $n \in N$ if $n \geq m$ then $\gamma(n) \geq r$, is a *test pair* for the sequence γ.

Theorem 16. A Cantor sequence γ is positive if and only if it has a test pair.

proof: Definitions 17 and 18.∗

The next three theorems show that the concept of positiveness introduced in Definition 17 agrees with some of our intuitive notions.

* Theorem 17. If γ and η are positive Cantor sequences, then so are $\gamma + \eta$ and $\gamma \cdot \eta$.

proof: Since γ and η are positive they have test pairs (k,r), $(l,s) \in N \times R^+$, so that

$$\text{for all } n \geq k, \; \gamma(n) \geq r, \text{ and}$$
$$\text{for all } n \geq l, \; \eta(n) \geq s.$$

(a) To prove that $\gamma + \eta$ is positive let $m = \max \{k,l\}$. We claim that $(m, r + s)$ is a test pair for $\gamma + \eta$. In fact, if $n \geq m$ then $n \geq k$ and $n \geq l$, whence

$$(\gamma + \eta)(n) = \gamma(n) + \eta(n) \geq r + s.$$

(b) With the same notation as above, we assert that (m,rs) is a test pair for $\gamma \cdot \eta$, therefore $\gamma \cdot \eta$ is positive. Indeed, if $n \geq m$ then

$$(\gamma \cdot \eta)(n) = \gamma(n) \cdot \eta(n) \geq r \cdot s,$$

whence the statement follows.

q.e.d.

Definition 19. A Cantor sequence γ is *negative* if and only if $-\gamma$ is positive.

Theorem 18. (Principle of Trichotomy.) If γ is a Cantor sequence, then one and only one of the following holds:

   (i) γ is positive;
   (ii) γ is a null sequence;
   (iii) γ is negative.

proof: It suffices to establish the following facts:

   (a) A null sequence is neither positive nor negative;
   (b) A sequence cannot be both positive and negative;
   (c) If a Cantor sequence is neither positive nor negative then it is null.

From (a) and (b) it follows that at most one of the alternatives (i), (ii), (iii) holds. For, if any two held, we would have a contradiction of (a) or of (b). From (c) we deduce that at least one of (i), (ii), (iii) holds.

(a) We prove first that if γ is null then γ is not positive. Suppose, to the contrary, that $\gamma \in \mathfrak{N}$ and γ is positive. Then there exists a nullity test $\varphi : R^+ \longrightarrow N_1$ (Definition 12) and also a test pair (m,r) for γ (Theorem 16). Let n be a natural number such that $n \geq \max \{m, \varphi(r/2)\}$. Then, on the one hand, $\sigma(n) \geq r$. But, on the other hand, $|\sigma(n)| = \sigma(n) < r/2$, a contradiction.

Also, $\gamma \in \mathfrak{N}$ implies that γ is not negative. Indeed, if γ were negative then $-\gamma$ would be positive. But $\gamma \in \mathfrak{N}$ also implies $-\gamma \in \mathfrak{N}$ (Exercise 4, page 196). Thus, $\gamma \in \mathfrak{N}$ and γ negative yield a sequence which is both null and positive. This has been proved impossible. Therefore, if $\gamma \in \mathfrak{N}$, then γ cannot be negative.

(b) Suppose a Cantor sequence γ is both positive and negative. Then γ and $-\gamma$ are both positive. By Theorem 17, $\gamma + (-\gamma)$ is positive, and by Theorem 3, $\gamma + (-\gamma) = \hat{0} \in \mathfrak{N}$, contrary to part (a).

(c) Suppose the Cantor sequence γ is neither positive nor negative. Since γ is not negative, $-\gamma$ is not positive. Therefore, the assumptions imply that neither γ nor $-\gamma$ is positive. We prove that $\gamma \in \mathfrak{N}$, and to this end define a nullity test ψ for γ.

Since $\gamma \in \mathfrak{C}$ it has a convergence test φ. Define ψ by

$$\psi(r) = \varphi\left(\frac{r}{2}\right),$$

for all $r \in R^+$. Since γ is not positive it has no test pair. In particular, if s is any positive rational number, $(\psi(s), s/2)$ is not a test pair. Hence, there exists a natural number $n_1 \geq \psi(s)$ such that $\gamma(n_1) < s/2$. Further, if $n \geq \psi(s)$ then

$$\gamma(n) = \gamma(n_1) + (\gamma(n) - \gamma(n_1))$$
$$\leq \gamma(n_1) + |\gamma(n) - \gamma(n_1)|.$$

But $n, n_1 \geq \psi(s)$ yield $n, n_1 \geq \varphi(s/2)$; and, since $\gamma \in \mathfrak{C}$, it follows that $|\gamma(n) - \gamma(n_1)| < s/2$. Consequently,

$$\gamma(n) < \frac{s}{2} + \frac{s}{2} = s.$$

Finally (as in Theorem 8), since φ is a convergence test for γ, it is also a convergence test for $-\gamma$. Thus, $\gamma(n) < s$ yields $-\gamma(n) < s$. And from these two inequalities we deduce $|\gamma(n)| < s$. In short, for any positive rational number s, if $n \geq \psi(s)$ then $|\gamma(n)| < s$, i.e., γ is a null sequence.

q.e.d.

Theorem 19. If γ is a positive (negative) sequence and if ν is a null sequence then $\gamma + \nu$ is positive (negative).

proof: Let $\rho = \gamma + \nu$. If ρ is null then from $\gamma = \rho - \nu$ it follows that γ is null. But by hypothesis γ is positive; hence, by Theorem 18, γ cannot be null, therefore neither is ρ. If ρ is negative then from $\nu = (-\gamma) + \rho$ we deduce ν is negative. This, too, is a contradiction. Hence, ρ is neither null nor negative, therefore by Theorem 18, ρ is positive.

The second part of the theorem is proved similarly.

q.e.d.

corollary: If γ is positive (negative) and if $\eta \in \bar{\gamma}$ then η is also positive (negative).

proof: We consider only the case that γ is positive. Since $\eta \in \bar{\gamma}$, it follows that $\eta - \gamma = \nu$ is a null sequence. Hence, $\eta = \gamma + \nu$, and by Theorem 19, η is positive.

q.e.d._*

The reader should notice that the order concept for Cantor sequences differs in a striking way from the corresponding concept for the rationals. For the rationals we have an order relation $>$ for which the following statement is true: for all a, $b \in R$ one and only one of the relations $a > b$, $a = b$, or $b > a$ holds. We could try defining $\alpha > \beta$ ("the sequence α is greater than the sequence β") by: $\alpha > \beta$ means $\alpha - \beta$ is positive. But for this relation we cannot prove that exactly one of $\alpha > \beta$, $\alpha = \beta$ or $\beta > \alpha$ holds. Indeed, as the following example shows, it is possible for none of these relations to hold for a pair of Cantor sequences:

Let α be the sequence defined by $\alpha(n) = 1/2^n$, $n \in N_1$,
β the sequence defined by $\beta(n) = -1/2^n$, $n \in N_1$.

Clearly, $\alpha \neq \beta$; but $\alpha > \beta$ and $\beta > \alpha$ are both false.

This difficulty will not arise when we use the concept of positiveness for Cantor sequences to define the corresponding concept for the real numbers.

* Definition 20. Let $u \in R$; then there is an $\alpha \in \mathcal{C}$ such that $u = \bar{\alpha}$. u is *positive* if and only if α is positive.

Suppose $u = \bar{\alpha} = \bar{\beta}$; by the corollary to Theorem 19 we know that β is positive if and only if α is. Hence, the definition of positiveness for a real number is independent of the particular sequence representing the equivalence class, that is, the given real number.

Theorem 20. Sums and products of positive real numbers are positive.

proof: Let u, v be positive real numbers. Then there exist positive Cantor sequences α and β such that $u = \bar{\alpha}$, $v = \bar{\beta}$.

(a) $u \oplus v = \bar{\alpha} \oplus \bar{\beta} = \overline{\alpha + \beta}$; but $\alpha + \beta$ is a positive Cantor sequence by Theorem 17. Hence, $\alpha + \beta$ is a positive real number.

(b) Similarly, from the fact that if α, β are positive Cantor sequences so is $\alpha \cdot \beta$ (Theorem 17), it follows that $u \odot v = \bar{\alpha} \odot \bar{\beta} = \overline{\alpha \cdot \beta}$ is positive.

q.e.d.

Theorem 21. If u is a real number then one and only one of the following holds:

  (i) u is positive;
  (ii) $u = \bar{0}$;
  (iii) $-u$ is positive.

proof: Since u is a real number, $u = \bar{\alpha}$ for some Cantor sequence α. By Theorem 18, exactly one of: α is positive, α is null, $-\alpha$ is positive, holds. If α is positive, so is u; if $-\alpha$ is positive, so is $-u$; if α is null, then $\bar{\alpha} = \bar{0} = u$. Since exactly one of these alternatives holds for α, the theorem is proved.

q.e.d.

Definition 21. If u is a real number such that $-u$ is positive, then u is *negative*. R^+ is the set of positive real numbers; R^- is the set of negative real numbers.∗

It is now possible to define an order relation "greater than" on the set R of real numbers by means of

∗ **Definition 22.** For all u, $v \in R$, u *is greater than* v (written "$u > v$") if and only if $u - v \in R^+$. u *is less than* v (written "$u < v$") if and only if $u - v \in R^-$.∗

The difficulty mentioned in connection with an order relation for Cantor sequences (defined in terms of positiveness for these sequences) does not arise for the real numbers. In fact, an immediate consequence of Theorem 21 and Definition 21 is the

∗**corollary:** For all real numbers u, v one and only one of $u > v$, $u = v$, $v > u$ holds.

Theorem 22. Let u, v, w, x be real numbers.

  (a_1) If $u > v$ then $u \oplus w > v \oplus w$;
  (b_1) if $u > v$ and $w > x$ then $u \oplus w > v \oplus x$;
  (a_2) if $u > v$ and $w > \bar{0}$ then $u \odot w > v \odot w$;
  (b_2) if $u > v$ and $w > x > \bar{0}$ then $u \odot w > v \odot x$;
  (a_3) if $u > v$ and $\bar{0} > w$ then $v \odot w > u \odot w$;
  (b_3) if $u > v$ and $\bar{0} > w > x$ then $v \odot x > u \odot w$;
  (c_1) if $u > \bar{0}$ and $v > \bar{0}$ then $u \odot v > \bar{0}$;
  (c_2) if $u > \bar{0}$ and $\bar{0} > v$ then $\bar{0} > u \odot v$;
  (c_3) if $\bar{0} > u$ and $\bar{0} > v$ then $u \odot v > \bar{0}$.

proof: (a_1) $(u \oplus w) - (v \oplus w) = u - v \in R^+$.

(b_1) $u \oplus w > v \oplus w > v \oplus x$.

(a_2) $(u \odot w) - (v \odot w) = (u - v) \odot w$.   Since $u - v$ and
$w \in R^+$, $(u \odot w) - (v \odot w) \in R^+$, whence $u \odot w > v \odot w$.

(b_2) Use (a_2).

(a_3) $(v \odot w) - (u \odot w) = (v - u) \odot w = (u - v) \odot (-w)$.
Since $u - v$ and $-w \in R^+$, $(u - v) \odot (-w) \in R^+$, it follows
that $v \odot w > u \odot w$.

(b_3), (c_1), (c_2) and (c_3) follow easily from the above.

q.e.d.∗

The elementary arithmetic of R will now be completed by proving
that every element $u \in R$, $u \neq \bar{0}$, has a "reciprocal" or "multiplicative
inverse" in R, from which it will follow easily that R is a field (see
Definition 29, Chapter 3).

∗lemma: Suppose α is a sequence such that

there exists an $n \in N_1$ such that for all natural numbers
$n \geq m$, $\alpha(n) = 0$.

Then α is a null sequence.

proof: The lemma is proved by exhibiting a nullity test $\psi : R^+ \longrightarrow N_1$
for α.   Define ψ by

$$\psi(r) = m \text{ for all } r \in R^+.$$

Then, if $n \geq \psi(r) = m$ it follows that $|\alpha(n)| = |0| = 0 < r$, hence,
ψ is indeed a nullity test for α.

q.e.d.

Theorem 23.   Let α be a positive sequence, (m, r) a test pair for α.   Define
a sequence $\beta : N_1 \longrightarrow R$ by

$$\beta(n) = \begin{cases} 0, & \text{if } n < m, \\ (\alpha(n))^{-1}, & \text{if } n \geq m. \end{cases}$$

Then

(a) β is a Cantor sequence, and

(b) $\alpha \cdot \beta \in \bar{1}$.

proof: (a) Let $\varphi: R^+ \longrightarrow N_1$ be a convergence test for α. Using φ we construct a convergence test $\psi: R^+ \longrightarrow N_1$ for β, whence β is a Cantor sequence.

Define ψ by

$$\psi(s) = \max \{m, \varphi(s \cdot r^2)\}, \ s \in R^+.$$

Let $n_1, n_2 \geq \psi(s)$; then $n_1, n_2 \geq m$ and therefore $\alpha(n_1) \geq r > 0$, $\alpha(n_2) \geq r > 0$, hence

$$| \beta(n_1) - \beta(n_2) | = \left| \frac{1}{\alpha(n_1)} - \frac{1}{\alpha(n_2)} \right|$$

$$= \left| \frac{\alpha(n_2) - \alpha(n_1)}{\alpha(n_1)\alpha(n_2)} \right|$$

$$= \frac{| \alpha(n_2) - \alpha(n_1) |}{\alpha(n_1)\alpha(n_2)}$$

$$\leq \frac{| \alpha(n_2) - \alpha(n_1) |}{r^2}.$$

But, $n_1, n_2 \geq \psi(s)$ also yield $n_1, n_2 \geq \varphi(s \cdot r^2)$; hence, $| \alpha(n_2) - \alpha(n_1) | < s \cdot r^2$.

Consequently,

$$| \beta(n_1) - \beta(n_2) | < \frac{s \cdot r^2}{r^2} = s,$$

and ψ is a convergence test for β.

(b) The sequence $\alpha \cdot \beta - \hat{1}$ obviously satisfies the condition

$$(\alpha \cdot \beta - \hat{1})(n) = 0$$

for all $n \geq m$. Hence, by the lemma, $\alpha \cdot \beta - \hat{1}$ is a null sequence, therefore $\alpha \cdot \beta \in \bar{1}$.

q.e.d.

Definition 23. Let $u \neq \bar{0}$ be an element in R. An element $v \in R$ such that $u \odot v = \bar{1}$ is a *reciprocal* or *multiplicative inverse* of u.

Theorem 24. Every element $u \in R$, $u \neq \bar{0}$ has a reciprocal $v \in R$.

proof: Let $u \in R$, $u \neq \bar{0}$. Then there is a Cantor sequence α such that $\bar{\alpha} = u$ and α is not a null sequence. If α is positive, then by Theorem 23 there is a Cantor sequence β such that $\alpha \cdot \beta = \hat{1}$. Since β is a Cantor sequence it determines a real number $v = \bar{\beta}$ and

$$\bar{1} = \overline{\alpha \cdot \beta} = \bar{\alpha} \odot \bar{\beta} = u \odot v,$$

so that v is a reciprocal of u.

If α is negative then $-\alpha$ is positive and there is a Cantor sequence, say $-\beta$, such that $(-\alpha) \cdot (-\beta) = \hat{1}$. Then with $v = \overline{-\beta}$ we have

$$\bar{1} = \overline{(-\alpha) \cdot (-\beta)} = \overline{(-\alpha)} \odot \overline{(-\beta)} = u \odot v,$$

and v is a reciprocal of u.

q.e.d._*

EXERCISES

1. If $u, v, w \in R_f$, $u \neq \bar{0}$, and $u \odot v = u \odot w$, then $v = w$. (Hence, R_f is an integral domain.   See Definition 28, Chapter 3).

2. If $u, w \in R_f$, $u \neq \bar{0}$, then there is one and only one $z \in R_f$, such that $u \odot z = w$. (Hence, R_f is a field.   Note the corollary: a nonzero element in R_f has exactly one reciprocal in R_f).

3. Prove: If $u \in R_f$, $u \neq \bar{0}$, then $u^2 = u \odot u$ is positive.

5

THE DEEPER STUDY OF
THE REAL NUMBERS

5.1. ORDERED FIELDS

In constructing successively the systems N, Z and R, it was proved that

there is an isomorphism between N and $\mathfrak{N} = \{ x \mid x \in Z \text{ and } x \geq 0 \}$;
there is an isomorphism between Z and $Z^* = \{ x/1 \mid x \in Z \}$.

This leads to the conjecture that there is an isomorphism between R and some subset of $\mathbb{R}$. Such a result can be established. It will be obtained in Section 5.2 as a corollary of the more general theorem:

Every ordered field K contains a subfield which is order-isomorphic with R.

Our first steps are to define "isomorphism" for fields, "ordered field," "order-preserving" isomorphism, and to deduce some important properties of ordered fields.

* Definition 1. Let K be a field with binary operations $+$ and $\cdot$, and let G be a set with a pair of binary operations $\oplus$ and $\odot$. An *isomorphism* between K and G is a one-one correspondence f between K and G such that:

For all $x, y \in K$,

(1) $f(x + y) = f(x) \oplus f(y)$, and
(2) $f(x \cdot y) = f(x) \odot f(y)$.

If there exists an isomorphism between K and G, then K and G are *isomorphic to each other*, and G is the *isomorphic image of K under f*.

Theorem 1. Let K, G and f be as in Definition 1. Then G is a field with respect to the operations $\oplus$ and $\odot$. (Theorem 1 is frequently stated in the form: An isomorphic image of a field is a field.)

proof: This is simply a matter of verifying that G, $\oplus$ and $\odot$ satisfy the conditions for a field (see Definition 29, Chapter 3). We illustrate with a few of the conditions and leave the rest to the reader.

Addition in G is commutative. For, if $u, v \in G$, then since f is one-one and onto, there exist $x, y \in K$ such that $f(x) = u, f(y) = v$. Then,

$$\begin{aligned}
u \oplus v = f(x) \oplus f(y) &= f(x + y) && \text{((1) of Definition 1)}\\
&= f(y + x) && \text{(Commutativity of } + \text{ in } K)\\
&= f(y) \oplus f(x) && \text{((1) of Definition 1)}\\
&= v \oplus u.
\end{aligned}$$

If 0 is the zero element of K then $f(0)$ is a zero element of G. We prove that for all $u \in G$, $u \oplus f(0) = u$. But, if $u \in G$, then there is an $x \in K$ such that $f(x) = u$. Then,

$$\begin{aligned}
u \oplus f(0) &= f(x) \oplus f(0)\\
&= f(x + 0)\\
&= f(x)\\
&= u.
\end{aligned}$$

If 1 is the unity element in K, then $f(1)$ is a unity element in G. Indeed, for each $u \in G$ there is an $x \in K$ such that $f(x) = u$. Then,

$$u \odot f(1) = f(x) \odot f(1) = f(x \cdot 1) = f(x) = u.$$

Every nonzero element $u \in G$ has a reciprocal in G. For, let $x \in K$, $f(x) = u$. Since f is one-one and u is not the zero element of G, $x \neq 0$, hence x has a reciprocal x^{-1} in K. Therefore,

$$f(1) = f(x \cdot x^{-1}) = f(x) \odot f(x^{-1}) = u \odot f(x^{-1})$$

so that $f(x^{-1})$ is a reciprocal of u.

The remaining conditions are verified in a similar way.

(Note: Since G is a field, $f(0)$ is *the* zero element of G and $f(1)$ is *the* unity element of G; and each nonzero element in G has exactly one reciprocal in G.)

q.e.d.∗

EXERCISE

Let f be an isomorphism between the fields K and G. Prove that the inverse mapping f^{-1} is an isomorphism between G and K.

* **Definition 2.** Let K be a field with binary operations $+$ and $\cdot$. K is an *ordered field with respect to a subset P* if and only if

(i) for each $x \in K$, one and only one of

$$x \in P, \ x = 0, \ -x \in P,$$

holds;

(ii) for all $x, y \in P$ both $x + y$ and $x \cdot y \in P$.

P is a set of *positive* elements of K if and only if K is an ordered field with respect to P; in this case, $K - \{0\} - P$ is the corresponding set of *negative* elements of K.∗

remarks

1. Since a field contains at least two elements it follows from (i) that $P \neq \phi$.

2. A field may be ordered with respect to distinct subsets (see Example, page 243 below).

3. The field R is ordered with respect to exactly one subset (see Corollary to Theorem 7, page 242).

EXERCISES

If the field K is ordered with respect to a subset P, then

1. for each $x \in K$, $x \neq 0$, $x^2 = x \cdot x \in P$; and

2. the unity element of K is in P.

The next theorem characterizes ordered fields in familiar terms.

* Theorem 2.   A field K is an ordered field with respect to a subset P if and only if there is a binary relation $<$ on K such that
(a) for all x, $y \in K$ one and only one of

$$x < y, \ x = y, \ y < x,$$

holds;

(b) for all x, y, $z \in K$, if $x < y$ and $y < z$, then $x < z$;
(c) for all x, y, $z \in K$, if $x < y$, then $x + z < y + z$;
(d) for all x, y, $z \in K$, if $x < y$ and $0 < z$, then $xz < yz$.

Moreover, the set P with respect to which K is ordered is the set

$$P = \{x \mid x \in K \text{ and } 0 < x\}.$$

proof: Suppose, first, K is an ordered field, and let P be the subset of K consisting of all the positive elements in K.   For all x, $y \in K$ define

(5.1) $x < y$ if and only if $y - x \in P$.

We claim that the binary relation $<$ satisfies (a)–(d).

(a) If x, $y \in K$, then $y - x \in K$ and, by Definition 2, exactly one of
$$y - x \in P, \ y - x = 0, \ -(y - x) \in P,$$
holds.   In the first case, $x < y$; in the second case, $x = y$; and in the third, $y < x$.   Clearly, the three cases are mutually exclusive.
(b) Suppose $x < y$ and $y < z$.   This means that $y - x \in P$ and $z - y \in P$.   By (ii) of Definition 2, $(y - x) + (z - y) = z - x \in P$, whence $x < z$.
(c) Let $x < y$; then $y - x \in P$.   But for each $z \in K$, $y - x = (y + z) - (x + z)$.   Hence, $x + z < y + z$.
(d) Let $x < y$ and let $0 < z$.   Then $y - x \in P$ and $z = z - 0 \in P$.   By (ii) of Definition 2, $(y - x)z = yz - xz \in P$, therefore $yz < xz$.

Conversely, let $<$ be an order relation on K such that (a)–(d) hold; we show that

$$P = \{x \mid x \in K \text{ and } 0 < x\}$$

satisfies (i) and (ii) of Definition 2.   It is an easy exercise (for the reader) to prove that (i) holds.   Next, let $y \in P$ and $z \in P$.   Then $0 < y$ and $0 < z$.   By (c), $z = 0 + z < y + z$; and since $0 < z$ and $z < y + z$, we have $0 < y + z$, by (b). Therefore, $y + z \in P$. Finally, $0 < y$ and $0 < z$ yield $0 = 0 \cdot z < y \cdot z$ (by (d)), whence $y \cdot z \in P$. Therefore, (ii) holds.

q.e.d.

Definition 3. The binary relation $<$, defined by (5.1), is the relation *less than, corresponding to the set* P; "$x < y$" is read, "x is less than y."*

remarks

1. If K is an ordered field with respect to a subset P_1 and also with respect to a subset P_2, then corresponding to P_1 there is a binary relation, say $<_1$, satisfying (a)–(d), and corresponding to P_2 there is also a binary relation, $<_2$, satisfying the same conditions.

2. In discussing ordered fields, one ordinarily does not mention the subset of positive elements. Thus, a reference to an "ordered field K" is taken to mean:

$$K \text{ is an ordered field with respect to some subset } P$$

and the subset P is not specified. Also, if $<$ is the relation, less than, corresponding to P, the expression "corresponding to P" is usually omitted where no confusion can arise.

EXAMPLES

1. The field R, of rationals, is an ordered field. (See Theorem 29, Chapter 3, and Exercises, page 169.)

2. The field of real numbers R is an ordered field. (See Theorems 20 and 21, Chapter 4.)

EXERCISES

Let K be an ordered field with respect to P. Define

$$x > y \text{ if and only if } x - y \in P.$$

$x > y$ is read, "x is greater than y." Prove:

1. For all $x, y \in K$, one and only one of $x > y$, $x = y$, $y > x$ holds.

2. $>$ is transitive.

3. If $x > y$, then for each $z \in K$, $x + z > y + z$.

4. If $x > y$ and $0 > z$, then $z \cdot y > z \cdot x$.

5. Give an alternative definition of $>$ (in terms of $<$) and show that the properties of Exercises 1–4 are deducible from the new definition.

* Definition 4. Let K and G be ordered fields with binary relations (less than) $\underset{K}{<}$ and $\underset{G}{<}$ respectively. An isomorphism h between K and G such that

$$\text{for all } x, y \in F, \text{ if } x \underset{K}{<} y \text{ then } h(x) \underset{G}{<} h(y),$$

is an *order-preserving* isomorphism. K and G are *order-isomorphic* if and only if there is an order-preserving isomorphism between them.

Theorem 3. Let K and G be ordered fields with binary relations $\underset{K}{<}$ and $\underset{G}{<}$, respectively, and let P_1 and P_2 be the corresponding sets of positive elements. If η is an isomorphism between K and G such that

$$P_2 = \{\, \eta(x) \mid x \in P_1 \},$$

then η is an order-preserving isomorphism between K and G.

proof: It suffices to show that for all $x, y \in K$, if $x \underset{K}{<} y$ then

$$\eta(x) \underset{G}{<} \eta(y).$$

Since $x \underset{K}{<} y$, $0 \underset{K}{<} y - x$, hence $y - x \in P_1$ and therefore $\eta(y - x) \in P_2$. But, since η is an isomorphism,

$$\eta(y - x) = \eta(y) - \eta(x),$$

whence

$$\eta(y) - \eta(x) \in P_2,$$

and

$$\eta(x) \underset{G}{<} \eta(y).$$

q.e.d.

corollary: Let K be an ordered field with respect to P_1 and let G be a field. If σ is an isomorphism between K and G, then G is an ordered field with respect to

$$P_2 = \{\sigma(x) \mid x \in P_1\},$$

and σ is an order-preserving isomorphism.

This Corollary is usually abbreviated by the statement:

An isomorphic image of an ordered field is an ordered field.

proof: We show that P_2 satisfies the conditions of Definition 2. It then follows, by Theorem 3, that σ is an order-preserving isomorphism between K and G.

(i) Let $u \in G$; there is a unique $x \in K$ such that $\sigma(x) = u$. If $u = 0$ then $x = 0$ (why?) and since $0 \notin P_1$ it follows that $u = 0 = \sigma(0) \notin P_2$. If $u \neq 0$ then $x \neq 0$ and exactly one of

$$x \in P_1, \ -x \in P_1,$$

holds. In the first case, $u = \sigma(x) \in P_2$, and in the second, $-u = -\sigma(x) = \sigma(-x) \in P_2$. Clearly, $u \in P_2$ and $-u \in P_2$ cannot both hold.

(ii) Let $+$ and $\cdot$ be the binary operations in K and let $\oplus$ and $\odot$ be the binary operations in G.

Suppose $u, v \in P_2$. Then there exist unique elements $x, y \in P_1$ such that $\sigma(x) = u$ and $\sigma(y) = v$. Since K is ordered with respect to P_1, $x + y$ and $x \cdot y \in P_1$. Hence, $\sigma(x + y) = \sigma(x) \oplus \sigma(y) = u \oplus v \in P_2$ and $\sigma(x \cdot y) = \sigma(x) \odot \sigma(y) = u \odot v \in P_2$.

Thus G is an ordered field with respect to P_2, and the corollary is proved.

q.e.d.∗

EXERCISES

1. Prove that the inverse of an order-preserving isomorphism between fields is order-preserving.

2. Prove that a composite of order-preserving isomorphisms is an order-preserving isomorphism.

5.2. RELATIONS BETWEEN ORDERED FIELDS AND R, THE FIELD OF RATIONAL NUMBERS

* **Definition 5.** Let K be a field with operations $+$ and $\cdot$, and let F be a subset of K. Further, let $\oplus$ and $\odot$ be the restrictions of $+$ and $\cdot$, respectively, to F. If F is a field with respect to $\oplus$ and $\odot$, then F is a *subfield* of K.∗

Examples of subfields will be given later.

The work required to show that (according to Definition 5) a subset of a field is a subfield, is a bit tedious. The next theorem gives a criterion which is easier to apply.

* **Theorem 4.** Let K be a field (with binary operations $+$ and $\cdot$) and let F be a subset of K such that

  (i) there is a nonzero element in F;
  (ii) for all $x, y \in F$, $x - y = x + (-y) \in F$ and $x \cdot y \in F$;
  (iii) for each $x \in F$, if $x \neq 0$, then $x^{-1} \in F$.

Then F is a subfield of K.

proof: As immediate consequences of the hypotheses we have:

(a) the zero element, 0, of K is in F. For, since there is an $x \in F$, by (ii), $0 = x - x \in F$.

(b) the unity element, 1, of K is in F. By (i) there is a nonzero x in F. By (iii), $x^{-1} \in F$. By the second part of (ii), $1 = x \cdot x^{-1} \in F$.

(c) if $x \in F$, then also $-x \in F$. By (a), $0 \in F$, hence by (ii), $-x = 0 - x \in F$.

(d) the restriction, $\oplus$, of $+$ to F is a binary operation on F. We prove that if $x, y \in F$, then $x \oplus y \in F$. But, if $x, y \in F$, then x and $-y \in F$, and by the first part of (ii), $x - (-y) \in F$. Since

$$x \oplus y = x + y = x - (-y),$$

$x \oplus y \in F$.

(e) the restriction, $\odot$, of $\cdot$ to F is a binary operation on F. Here we show that if $x, y \in F$, then $x \odot y \in F$. But $x \odot y = x \cdot y$ (definition of restriction) and by the second part of (ii), $x \cdot y \in F$. Hence, $x \odot y \in F$.

All that remains now is to show that with $\oplus$ and $\odot$ as binary operations, F is a field. We verify a few of the conditions and leave the rest as exercises.

Commutative Law for $\oplus$.

$$
\begin{aligned}
x \oplus y &= x + y && \text{(Definition of } \oplus) \\
&= y + x && \text{(Commutativity of } + \text{ in } K) \\
&= y \oplus x && \text{(Definition of } \oplus).
\end{aligned}
$$

Distributive Law.

$$
\begin{aligned}
x \odot (y \oplus z) &= x(y + z) && \text{(Definitions of } \oplus \text{ and } \odot) \\
&= xy + xz && \text{(Distributive law in } K) \\
&= x \odot y \oplus x \odot z && \text{(Definitions of } \oplus \text{ and } \odot).
\end{aligned}
$$

q.e.d.∗

The main theorem of this section is the one referred to at the beginning of 5.1.

∗ Theorem 5. Let K be an ordered field. Then K contains a subfield R' such that R (the rationals) and R' are order-isomorphic.

proof: We construct a mapping $h: R \longrightarrow K$ of R into K in several steps, and show that h is an order-preserving isomorphism.

Let $0'$, $1'$ be the zero and unity elements, respectively, of K, and let $\oplus$, $\odot$ be the binary operations on K. Set

$$
h(0) = 0', \; h(1) = 1',
$$

where $0, 1 \in R$. We define h by recursion on the nonnegative integers in R. Let m be any nonnegative integer and assume that $h(m)$ is defined; this means we are assuming there is one and only one element $m' \in K$ such that $h(m) = m'$. Set

$$
h(m + 1) = h(m) \oplus 1';
$$

then $h(n)$ has been defined recursively for all the nonnegative integers in R.

Although the definition of h is not yet complete, we can establish that h has several useful properties.

(a) For all positive integers, n, $h(n)$ is a positive element in K. Since $h(1) = 1'$, and since the unity element of an ordered field is positive, $h(1)$ is positive. Let m be any positive integer and assume $h(m)$ is positive. Then $h(m + 1) = h(m) \oplus 1'$, and since $h(m)$ and $1'$ are positive, the properties of an ordered field assure us that $h(m) \oplus 1'$ is positive. By the PFI, $h(n)$ is positive for all positive integers $n \in R$.

(b) For all nonnegative integers n, m in R, $h(n + m) = h(n) \oplus h(m)$. Again we use induction.   Let

> $L = \{x \mid x$ is a nonnegative integer such that for each nonnegative integer n, $h(n + x) = h(n) \oplus h(x)\}$.

By the definition of h, 0 and $1 \in L$.   Suppose $y \in L$; then

$$
\begin{aligned}
h(n + (y + 1)) &= h((n + y) + 1) \\
&= h(n + y) \oplus h(1) && \text{(Since } 1 \in L) \\
&= (h(n) \oplus h(y)) \oplus h(1) && (y \in L, \text{ by assumption)} \\
&= h(n) \oplus (h(y) \oplus h(1)) \\
&= h(n) \oplus h(y + 1) && \text{(Definition of } h).
\end{aligned}
$$

Hence, $h(n + m) = h(n) \oplus h(m)$ for all nonnegative integers n, $m \in R$.

(c) If n, m are distinct, nonnegative integers in R, then $h(n) \neq h(m)$. We may assume $n > m$, hence there is a positive integer p such that $n = m + p$. By (b), $h(n) = h(m + p) = h(m) \oplus h(p)$. Since $h(p)$ is positive (by (a)), $h(n) > h(m)$. And, since K is an ordered field, $h(n) \neq h(m)$.

(d) For all nonnegative integers n, $m \in R$, $h(n \cdot m) = h(n) \odot h(m)$. Let

> $M = \{x \mid x$ is a nonnegative integer in R such that for each nonnegative integer $n \in R$, $h(n \cdot x) = h(n) \odot h(x)\}$.

Clearly, 0, $1 \in M$.   Assume $y \in M$.   Then

$$
\begin{aligned}
h(n \cdot (y + 1)) &= h(n \cdot y + n \cdot 1) \\
&= h(n \cdot y) \oplus h(n \cdot 1) && \text{(By (b))} \\
&= h(n) \odot h(y) \oplus h(n) \odot h(1) \\
&= h(n) \odot (h(y) \oplus h(1)) && \text{(Distributive law in } K) \\
&= h(n) \odot h(y + 1) && \text{(Definition of } h).
\end{aligned}
$$

Therefore $h(n \cdot m) = h(n) \odot h(m)$ for all nonnegative integers n, m in R.
Next, we extend the definition of h by setting

$$h(x) = -h(-x)$$

for all negative integers $x \in R$. Let Z^* be the set of all integers in R. It is a simple chore (for the reader) to prove:

> h is an order-preserving mapping of Z^* into K;
> h is one-one;
> for all x, $y \in Z^*$,
> $\qquad h(x + y) = h(x) \oplus h(y)$ and $h(x \cdot y) = h(x) \odot h(y)$.

Finally, we complete the definition of h by setting

$$h\left(\frac{a}{b}\right) = h(a) \odot h(b)^{-1}, \quad \frac{a}{b} \in R, \quad a, b \in Z^*, \quad b \neq 0.$$

Since $b \neq 0$, $h(b) \neq 0$, therefore $h(b)^{-1}$ exists in K. We claim

$$\frac{a}{b} = \frac{c}{d} \text{ if and only if } h\left(\frac{a}{b}\right) = h\left(\frac{c}{d}\right).$$

Indeed, if $a/b = c/d$, then $ad = bc$, whence (since ad, bc are integers in R), $h(a \cdot d) = h(b \cdot c)$. Therefore, $h(a) \odot h(d) = h(b) \odot h(c)$ and

$$h(a) \odot h(b)^{-1} = h(c) \odot h(d)^{-1},$$

or

$$h\left(\frac{a}{b}\right) = h\left(\frac{c}{d}\right),$$

by the definition of h. Since these steps are reversible, we deduce from $h(a/b) = h(c/d)$ that $a/b = c/d$. Therefore, h is one-one.

It is easy to check that

$$h\left(\frac{a}{b} + \frac{c}{d}\right) = h\left(\frac{a}{b}\right) \oplus h\left(\frac{c}{d}\right) \text{ and } h\left(\frac{a}{b} \cdot \frac{c}{d}\right) = h\left(\frac{a}{b}\right) \odot h\left(\frac{c}{d}\right);$$

since h is one-one it is an isomorphism. Moreover, since $a/b < c/d$ if and only if $ad < bc$ (we may assume a, b, c, d chosen so that $bd > 0$), it follows that h is an order-preserving isomorphism of R into K.

Now let $R' = \{x' \mid x' \in K$ and there is an $x \in R$ such that $h(x) = x'\}$. By the corollary to Theorem 3, R' is a field which is order-isomorphic with R. Since (i)–(iii) of Theorem 4 hold for R', R' is a subfield of K.

q.e.d.

corollary 1: The field R of real numbers contains a subfield R^* which is order-isomorphic with R.

proof: It has been shown (Theorems 20 and 21, Chapter 4) that R is an ordered field. Hence, taking $K = R$ in Theorem 5, we deduce that R contains a subfield R^* order-isomorphic with R.

q.e.d.

corollary 2: Let K be an ordered field and let R' be a subfield of K order-isomorphic with R. Then there exists an order-preserving isomorphism between R' and R^*.

proof: Let $g:R \longrightarrow R'$ be an order-preserving isomorphism between R and R' and let $f:R \longrightarrow R^*$ be an order-preserving isomorphism between R and R^*. Then g^{-1} is an order-preserving isomorphism between R' and R and $f \circ g^{-1}:R' \longrightarrow R^*$ is an order-preserving isomorphism between R' and R^* (see Exercises 1 and 2, page 220).

corollary 3: Let K_1 and K_2 be ordered fields. Then there exist subfields R_1' and R_2' of K_1 and K_2, respectively, such that R_1', R_2' and R (the field of rationals) are mutually order-isomorphic.

proof: Exercise.

q.e.d.

remark: It can be proved that an ordered field contains only one field R' order-isomorphic to R and that there is only one isomorphism between R and R'. The proof is not carried out here.

We now give a direct proof of Corollary 1 (above) which does not use Theorem 5. This proof exhibits the isomorphism f in a simple way. Corollary 1 is restated as follows:

Let $R^* = \{\bar{r} \mid r \in R\}$. Then the mapping f, defined by

$$f(r) = \bar{r}, r \in R,$$

is an order-preserving isomorphism between R and R^*.

lemma: If $r, s \in R$ then $\bar{r} = \bar{s}$ if and only if $r = s$.

proof: Clearly, if $r = s$ then $\bar{r} = \bar{s}$.

For the converse, it suffices to show that if $r \in R$ and $\bar{r} = \bar{0}$, then $r = 0$. For, if this statement is true, then from $\bar{r} = \bar{s}$ follows $\bar{0} = \bar{r} - \bar{s} = \overline{r - s}$. Hence, $r - s = 0$ and $r = s$.

To prove that if $\bar{r} = \bar{0}$, where $r \in R$, then $r = 0$, we need only observe that $\hat{r}$ is positive, null, or negative according as r is positive, zero, or negative, and $\hat{r} \in \bar{r}$.

q.e.d.   (Lemma)

Second Proof of Corollary 1: By the lemma, f is a one-one correspondence between R and R^*. Moreover, f satisfies the conditions of Definitions 1 and 4, hence is an order-preserving isomorphism. Indeed, for all $x, y \in R$,

(i) $\qquad f(x + y) = \overline{x + y} = \overline{x} \oplus \overline{y} = f(x) \oplus f(y);$

(ii) $\qquad f(x \cdot y) = \overline{x \cdot y} = \overline{x} \odot \overline{y} = f(x) \odot f(y);$

(iii) If $y < x$ then $x - y > 0$, hence $\overbrace{x - y}$ is a positive Cantor sequence. Therefore $\overline{x - y} = \overbrace{x - y}$ is a positive real number. But $\overline{x - y} = \overline{x} - \overline{y}$, so that $\overline{x} - \overline{y} > 0$, therefore $\overline{y} < \overline{x}$, i.e., $f(y) < f(x)$. This completes the second proof of Corollary 1.

$$\text{q.e.d.}_*$$

By the above Remark, the isomorphism f, defined in the second proof of Corollary 1, is the only isomorphism between R and R^*.

It is now time to clean up the notation for the real numbers. If $r \in R$, then under the isomorphism f (second proof of Corollary 1 to Theorem 5), $f(r) = \overline{r}$; henceforth the symbol "r" will denote the real number $\overline{r}(= \overline{\overline{r}})$. If u, v are real numbers, their sum $u \oplus v$ and their product $u \odot v$ will be denoted by "$u + v$" and "$u \cdot v$" (or, more commonly, uv) respectively. In most contexts, either it is clear that

"r" denotes the rational number r, or the real number $\overline{r}$;

"$+$" denotes sum in R or $\oplus$ in $\mathbb{R}$;

"$\cdot$" denotes product in R or $\odot$ in $\mathbb{R}$,

or else it does not much matter in which of the two ways these symbols are interpreted. Such distinctions are important only when we are concerned with the nature of number systems rather than with their use.

Let us review, briefly, the meanings of the symbols for numbers and binary operations.

A symbol such as "2" now has four meanings: it may denote an element of N, of Z, of R, or of $R^* \subset \mathbb{R}$. Whichever of the four systems we have in mind we usually speak of the "natural number, two" or "the integer, two."

A symbol such as "-2" has three meanings: it may denote an element of Z, of R, or of R^*. In any case, it is usually called "the integer, minus two."

A symbol such as "$\frac{2}{3}$" may denote an element of R or of R^* and is generally called "the fraction, two-thirds" or "the rational number, two-thirds."

A symbol such as "$\sqrt{2}$" can only denote an element of $\mathbb{R}$.

The field R^*, which is order-isomorphic with R and is a subset of $\mathbb{R}$, is frequently called "the field of rational numbers" even though, properly, this name should be reserved for R.

The symbols "$+$" and "$\cdot$" may denote operations in N, Z, R, R^*, or R and are called "plus" and "times" (or "sum" and "product") respectively.

Again we emphasize that in most cases either it is clear from the context which of the several systems is under consideration, or else it does not matter.

Note that a symbol such as "$a + b\sqrt{2}$," $a, b \in R$, stands for

$$\bar{a} \oplus (\bar{b} \odot \sqrt{2})$$

where the symbols "a," "b," "$\sqrt{2}$," "$\oplus$," "$\odot$" have the meanings discussed above.

Throughout the remainder of Chapter 5 it will be assumed that the generalized operations and the generalized commutative and associative laws have been defined for R in accordance with the discussion in Chapter 2 (Section 2.9). These concepts will be used as needed.

EXERCISES

1. Let S be the subset of R defined by $S = \{a + b\sqrt{2} \mid a, b \in R^*\}$. Prove that S is a subfield of R. Do the same for

    $$T = \{a + b\sqrt{5} \mid a, b \in R^*\}.$$

2. Let $U = \{a + b\sqrt[3]{2} \mid a, b \in R^*\}$. Is U a subfield of R? Is

    $V = \{a + b\sqrt[3]{2} + c(\sqrt[3]{2})^2 \mid a, b, c \in R^*\}$ a subfield of R?

5.3.   THE COMPLETENESS OF THE REAL NUMBERS

Although R and R are similar in that both are ordered fields, there are important differences between them. For instance, the number $2 \ (\in R)$ possesses no square root in R, whereas $2 \ (\in R)$ does have a square root in R (Example 2, pages 201–203, Chapter 4). From these facts, it is easy to deduce that:

There is no isomorphism between R and R. The proof of this statement is left as an exercise.

The major result of this section will be that R has the property of "completeness" to be defined below. From the completeness of R, we shall be able to prove (Section 5.4) that every nonnegative number in R has a square root in R. Further, we shall see (Section 5.6) that all ordered fields with the completeness property are order-isomorphic. Since we already know that R and $R_{\mathfrak{f}}$ are not isomorphic, it follows that R cannot be complete.

Because of the theorem to be proved in Section 5.6, we state several of our definitions and theorems for ordered fields. Since R and $R_{\mathfrak{f}}$ are instances of ordered fields, all of our results carry over, word for word, to these important special cases.

* Definition 6. Let $A \neq \phi$ be a subset of an ordered field K. An element $y \in K$ is an *upper bound* (abbreviated "u.b.") for A if and only if for each $x \in A$, $x \leq y$. If A has an upper bound, then A is *bounded above*.*

An upper bound for a set may or may not be an element of the set.

EXAMPLES

1. $K = R, A = \{x \mid x \in R \text{ and } 0 < x < 1\}$. Clearly, 17 is an u.b. for A as are $\frac{11}{3}$ and 1. For that matter, every rational number $r \geq 1$ is an u.b. for A, but no u.b. for A is an element of A.

2. $K = R, B = \{x \mid x \in R \text{ and } x \leq \frac{1}{2}\}$. Every rational number $\geq \frac{1}{2}$ is an u.b. for B, and in this case the u.b., $\frac{1}{2}$, is an element of B.

3. The set of all positive rationals has no u.b.

EXERCISES

1. Describe several subsets of R which possess upper bounds.

2. Prove: If a subset of an ordered field has an upper bound, then it has infinitely many.

3. (a) By analogy with Definition 6, define "lower bound" (abbreviated "l.b.") and "bounded below."
   (b) Give examples of subsets of ordered fields which possess lower bounds.
   (c) Prove: If a subset of an ordered field has one lower bound, then it has infinitely many.

4. Let $A \neq \phi$ be a subset of an ordered field, $B = \{x \mid -x \in A\}$. Prove that B has a lower bound if and only if A has an upper bound.

Now consider the subset

(5.2)
$$A = \left\{ 1 - \frac{1}{2^n} \mid n \in N_1 \right\}$$

of R. Obviously, 37, 15, 21, $\frac{3}{2}$, 1 are all upper bounds for A, but numbers smaller than one are not upper bounds. The reader can verify without too much trouble that even $1.\frac{9 \dot{9} \dot{9} \; 9 9 9 . 9 9 9}{0 0 0 . 0 0 0 , 0 0 0}$ (< 1) is not an upper bound for A. Thus one is, in a certain sense, a "best" upper bound for A. The concept of "best" upper bound is embodied in

* Definition 7. Let K be an ordered field, $A \neq \phi$ a subset of K. An element $y \in K$ is a *supremum* for A (abbreviated "sup A") if and only if

   (i) y is an u.b. for A, and
   (ii) if $z < y$ then z is not an u.b. for A.*

A supremum for a set A is also called a "least upper bound" for A. In general, we shall prefer the term "supremum" to "least upper bound."

EXAMPLES

1. Let A be the subset of R defined by (5.2). We claim that one is a supremum for A. Clearly, $1 \geq 1 - 1/2^n$ for each $n \in N_1$, so that condition (i) of Definition 7 is satisfied. Let $z < 1$; to verify that condition (ii) holds we must prove that z is not an u.b. for A. If $z \leq 0$, then obviously (ii) is satisfied, since all the elements of A are positive. Suppose $0 < z < 1$. It is easy to show (see footnote 3, Chapter 4) that there is a natural number m such that

$$2^m > \frac{1}{1 - z};$$

hence, by some elementary computation,

$$1 - \frac{1}{2^m} > z.$$

Therefore z is not an u.b. for A.

2. A subset of an ordered field may possess an upper bound, but *not* a supremum. Again, let the field be R and let

$$B = \{x \mid x \in R \text{ and } x \leq 0\} \cup \{x \mid x \in R \text{ and } x > 0 \text{ and } x^2 \leq 2\}.$$

B has an upper bound, for instance, two. However, B does not have a supremum in R. This result is an immediate consequence of the following:

Let $w > 1$ be a rational number.

(a) If $w^2 > 2$ then w is an u.b. for B.

(b) If $w^2 > 2$ then there is a rational number v, such that $v < w$ and $v^2 > 2$.

(c) If $w^2 < 2$ then there is a rational number u such that $w < u$ and $u^2 < 2$.

Assuming, for the moment, that (a), (b) and (c) have been proved, it is easy to show that B has no supremum in R. Indeed, suppose $z \in R$ is a supremum for B. Obviously, $z^2 \neq 2$ (Theorem 2, Chapter 4) and $z > 1$. If $z^2 < 2$, then by (c) there is an $s \in R$ such that $z < s$ and $s^2 < 2$. Hence, z is not an u.b. for B, therefore $z \neq \sup B$, a contradiction. On the other hand, if $z^2 > 2$, by (b) there is a rational number t such that $z > t$ and $t^2 > 2$. Therefore, t is an u.b. for B and again $z \neq \sup B$. Thus, in every case, the assumption that B has a supremum in R yields a contradiction.

All that remains is to prove statements (a), (b) and (c).

(a) The proof of (a) is simple and is left as an exercise.

(b) Note, first, that if r is a positive rational number, then there is a natural number $k > 0$ such that $2^k > r$ (why?). In particular, since $(w - 1)/(w^2 - 2) \in R$ and is positive there is a natural number $m - 1$ such that $2^{m-1} > (w - 1)/(w^2 - 2)$. By the arithmetic of the rationals

$$w^2 - 2 > \frac{w - 1}{2^{m-1}} = \frac{w}{2^{m-1}} - \frac{1}{2^{m-1}}.$$

Since $2^{2m} > 2^{m-1}$ for $m - 1 > 0$, $1/2^{m-1} < 1/2^{2m}$, whence

$$w^2 - 2 > \frac{w}{2^{m-1}} - \frac{1}{2^{2m}},$$

or

$$w^2 - \frac{2w}{2^m} + \frac{1}{2^{2m}} > 2.$$

Set $v = w - 1/2^m$; then $v < w$ and

$$v^2 = \left(w - \frac{1}{2^m}\right)^2 = w^2 - \frac{2w}{2^m} + \frac{1}{2^{2m}} > 2.$$

This proves (b).

(c) The proof of (c) follows that of (b), step by step, except that this time we have $(w - 1)/(2 - w^2) > 0$ and $2^{m-1} > (w - 1)/(2 - w^2)$. The details are left to the reader.

Thus, in the field R, it is possible for a subset to be bounded above yet to possess no supremum. By contrast, it will be seen that this situation cannot arise for $R̸$.

EXERCISES

1. Let $a \in R$ be a positive rational number such that for all $b \in R$, $a \neq b^2$. Prove that the set

   $$\{x \mid x \in R \text{ and } x \le 0\} \cup \{x \mid x \in R \text{ and } 0 < x \text{ and } x^2 \le a\}$$

   is bounded above but has no supremum in R.

2. Prove that a nonempty subset of an ordered field K has at most one supremum in K. This result justifies speaking of *the* supremum whenever a supremum exists.

* Definition 8. Let K be an ordered field, A a nonempty subset of K. An element $w \in K$ is an *infimum* for A (abbreviated "inf A") if and only if

   (i) w is a lower bound for A, and
   (ii) if $w < u$ then u is not a lower bound for A.*

Thus, an inf A is a "best" or "largest possible" lower bound for A.

EXERCISES

1. Prove that a nonempty subset of an ordered field K has at most one infimum. (Therefore we may speak of *the* infimum of A if an infimum exists.)

2. Let $A \neq \phi$ be a subset of an ordered field F, $B = \{x \mid -x \in A\}$. Prove that B has an infimum if and only if A has a supremum.

The important result which we want to prove is

* Theorem 6. Every nonempty subset A of $R̸$ which is bounded above has a supremum in $R̸$.

For purposes of clarity, the proof of Theorem 6 will be separated into a sequence of lemmas. Throughout the proof, f will be the isomorphism

$$f : R \longrightarrow R^* \subset R̸$$

defined in the second proof of Corollary 1 to Theorem 5, and as elsewhere in this chapter, if $r \in R$, then $\hat{r}$ is the sequence

$$\hat{r}(n) = r, \; n \in N_1,$$

where N_1 is the set $N - \{0\}$. For each $r \in R$, $f(r) = \bar{r} = \bar{\hat{r}}$ is the real number containing the Cantor sequence $\hat{r}$ as one of its elements.

Let η be a Cantor sequence. In the proofs of lemmas 1 and 2, it is important to distinguish among

$$\widehat{\eta(n)}, \; n \in N_1; \quad \overline{\eta(n)}, \; n \in N_1; \quad \text{and} \; \bar{\eta}.$$

To illustrate, consider the specific example:

$$\eta(n) = 1 - \frac{1}{2^n}, \quad n \in N_1.$$

η is a Cantor sequence and

$$\eta(1) = 1 - \frac{1}{2^1} = \frac{1}{2},$$

whence

$$\widehat{\eta(1)} = \left(\hat{\frac{1}{2}} \right).$$

Thus, $\widehat{\eta(1)}$ is the constant sequence

$$\widehat{\eta(1)}(n) = \left(\hat{\frac{1}{2}} \right)(n) = \frac{1}{2}, \quad n \in N_1.$$

Similarly,

$$\eta(2) = 1 - \frac{1}{2^2} = \frac{3}{4},$$

and

$$\widehat{\eta(2)}(n) = \left(\hat{\frac{3}{4}} \right)(n) = \frac{3}{4}, \quad n \in N_1,$$

etc. (What are $\widehat{\eta(3)}$ and $\widehat{\eta(4)}$?) By contrast, $\overline{\eta(1)} = \overline{\widehat{\eta(1)}}$ is the real number containing the (constant) Cantor sequence $\widehat{\eta(1)} = (1/2)$ as one of its elements. Hence,

$$\overline{\eta(1)} = \overline{\left(\frac{1}{2}\right)}.$$

Similarly, the reader can verify that

$$\overline{\eta(2)} = \overline{\left(\frac{3}{4}\right)}, \quad \overline{\eta(3)} = \overline{\left(\frac{7}{8}\right)}, \quad \overline{\eta(4)} = \overline{\left(\frac{15}{16}\right)}, \quad \text{etc.}$$

Finally, $\bar{\eta}$ is the real number containing the Cantor sequence η as one of its elements. Thus

$$\bar{\eta} = \bar{1}.$$

A fact that is especially needed is:

If η is a Cantor sequence, then for each $n \in N_1$,

(5.3) $$\widehat{\eta(n)}(k) = \eta(n), \ k \in N_1.$$

lemma 1: Let $w \in \mathbb{R}$ and let $m_1 \in N_1$. If σ is a Cantor sequence such that

$$\text{for all } n \in N_1, \ n \geq m_1, \ \overline{\sigma(n)} \geq w,$$

then $\bar{\sigma} \geq w$. (Important exercise: Construct an example to illustrate the meaning of the foregoing statement.)

proof: Since $w \in \mathbb{R}$, there is a Cantor sequence τ such that $w = \bar{\tau}$. Lemma 1 is proved by contradiction.

Suppose $\bar{\sigma} < \bar{\tau} = w$. Then $\tau - \sigma$ is a positive sequence, hence there is a test pair (m_2, r) such that for all $n \geq m_2$,

$$(\tau - \sigma)(n) = \tau(n) - \sigma(n) > r.$$

By (5.3), for all $n \in N_1, \widehat{\sigma(n)}(n) = \sigma(n)$ and so for all $n \geq \max\{m_1, m_2\}$,

$$\tau(n) - \widehat{\sigma(n)}(n) > r.$$

Since $\sigma \in \mathbb{C}$, σ has a convergence test φ. Therefore, if $m, n \geq \varphi(r/2)$,

$$|\sigma(n) - \sigma(m)| < \frac{r}{2},$$

whence

$$-\frac{r}{2} < \sigma(n) - \sigma(m).$$

Now let $m_3 = \max \{m_1, m_2, \varphi(r/2)\}$. Using (5.3) again we find that for all $n, m \geq m_3$,

$$\tau(n) - \widehat{\sigma(m)}(n) = \tau(n) - \widehat{\sigma(n)}(n) + \widehat{\sigma(n)}(n) - \widehat{\sigma(m)}(n)$$

$$= \tau(n) - \widehat{\sigma(n)}(n) + \sigma(n) - \sigma(m)$$

$$> r - \frac{r}{2} = \frac{r}{2}.$$

In brief,

$$(\tau - \widehat{\sigma(m)}) (n) > \frac{r}{2},$$

for all $n, m \geq m_3$. This means that $\tau - \widehat{\sigma(m)}$ is a positive Cantor sequence, hence

$$\overline{\tau - \widehat{\sigma(m)}} > \overline{0},$$

i.e.,

$$\bar{\tau} - \overline{\widehat{\sigma(m)}} = \bar{\tau} - \overline{\sigma(m)} > \overline{0},$$

or

$$w = \bar{\tau} > \overline{\sigma(m)},$$

for all $m \geq m_3$ where $m_3 \geq m_1$. This contradicts the hypothesis of the lemma. Therefore the assumption $w > \bar{\sigma}$ is false and Lemma 1 is proved.

lemma 2: Let $w \in \mathbb{R}$ and let $m_1 \in N_1$. If σ is a Cantor sequence such that

$$\text{for all } n \in N_1, n \geq m_1, \overline{\sigma(n)} \leq w,$$

then $\bar{\sigma} \leq w$.

proof: The proof of Lemma 2 is easily derived from that of Lemma 1 and is left to the reader.

lemma 3: Let $A \neq \phi$ be a subset of $\mathbb{R}$, and suppose A is bounded above. For each natural number n there is a unique integer k_n such that

$$\text{for all } x \in A, x < f\left(\frac{k_n}{2^n}\right),$$

and there is a $y \in A$ such that $y \geq f((k_n - 1)/2^n)$.

(*Note:* In the statement of Lemma 3, and elsewhere in the proof of Theorem 6, we refer to integers. These are considered as elements in R. Thus the integer k_n is to be regarded as the rational number $k_n/1$.)

proof: By the hypothesis of the lemma, A has an upper bound, say, u, and $u = \bar{\rho}$ where $\rho \in \mathcal{C}$. Since ρ is a Cantor sequence, it is bounded, hence there is an integer m such that

$$\rho(n) < m,$$

for all $n \in N_1$. Since $\rho(n)$ and $m \in R$, $f(\rho(n))$ and $f(m)$ are real numbers, and by the properties of f (f preserves order),

$$f(\rho(n)) < f(m).$$

Applying Lemma 2 with $\sigma = \rho$, $m_1 = 1$, $w = f(m)$, we have

$$u = \bar{\rho} \le f(m).$$

But, for all natural numbers a,

$$m < \frac{2^a \cdot m + 1}{2^a} \quad (\text{why?}).$$

Therefore,

$$f(m) < f\left(\frac{2^a \cdot m + 1}{2^a}\right),$$

for all $a \in N$.

Now let n be a natural number, and let k be an integer such that $k > 2^n \cdot m + 1$. Then $f(k/2^n) > f((2^n \cdot m + 1)/2^n)$. Hence, for all $x \in A$, $x \le u$ and therefore

(5.4)
$$x < f\left(\frac{k}{2^n}\right).$$

By a similar argument using Lemma 1, if $v \in A$, there is an integer t such that

(5.5)
$$v \ge f\left(\frac{t}{2^n}\right),$$

where n is the same natural number as in (5.4). It follows from (5.4) and (5.5) that there is a unique smallest integer k_n such that for all $x \in A$,

$$x < f\left(\frac{k_n}{2^n}\right).$$

Consequently, there is a $y \in A$ such that

$$y \geq f\left(\frac{k_n - 1}{2^n}\right).$$

This completes the proof of Lemma 3.

lemma 4: For each $n \in N_1$ let k_n be the integer determined in Lemma 3. Then the sequence $\sigma : N_1 \longrightarrow R$ defined by

$$\sigma(n) = \frac{k_n}{2^n}, \quad n \in N_1,$$

is a Cantor sequence.

proof: Clearly, σ is a function (why?). Thus we need only prove that σ has a convergence test.

If $n, m \in N_1$ with, say, $n \geq m$, then $k_n \leq 2^{n-m} k_m$. For,

$$\sigma(m) = \frac{k_m}{2^m} = \frac{2^{n-m} k_m}{2^n};$$

hence, for each $x \in A$,

$$x < f\left(\frac{k_m}{2^m}\right) = f\left(\frac{2^{n-m} k_m}{2^n}\right).$$

But k_n is the smallest integer such that for each $x \in A$, $x < f(k_n/2^n)$. Therefore $k_n \leq 2^{n-m} k_m$.

On the other hand, there is a $y \in A$ such that

$$y \geq f\left(\frac{k_m - 1}{2^m}\right) = f\left(\frac{2^{n-m} k_m - 2^{n-m}}{2^n}\right).$$

Consequently, $k_n \geq 2^{n-m} k_m - 2^{n-m}$. Therefore,

$$0 \leq 2^{n-m} k_m - k_n \leq 2^{n-m}.$$

Dividing by 2^n we have

(5.6)
$$0 \leq \frac{k_m}{2^m} - \frac{k_n}{2^n} \leq \frac{1}{2^m}.$$

Since $\sigma(m) = k_m/2^m$, $\sigma(n) = k_n/2^n$,

$$|\sigma(m) - \sigma(n)| \leq \frac{1}{2^m}.$$

We define a convergence test $\varphi : R^+ \longrightarrow N_1$ by setting

$$\varphi(r) = \text{smallest natural number } m \text{ such that } 1/2^m < r.$$

Clearly, φ is a convergence test for σ, and Lemma 4 is proved.

lemma 5:   Let σ and k_n, $n \in N_1$, be as in Lemma 4, and let $\bar{\sigma}$ be the real number determined by σ. Then, for each $m \in N_1$,

$$\bar{\sigma} \leq f\left(\frac{k_m}{2^m}\right).$$

proof: By (5.6), if n, $m \in N_1$, then for all $n \geq m$,

$$\sigma(n) = \frac{k_n}{2^n} \leq \frac{k_m}{2^m} = \sigma(m),$$

hence,

$$\overline{\sigma(n)} \leq \overline{\sigma(m)}.$$

Applying Lemma 2 with $w = \overline{\sigma(m)}$ we see that

$$\bar{\sigma} \leq w = \overline{\sigma(m)} = f\left(\frac{k_m}{2^m}\right),$$

hence Lemma 5 is proved.

lemma 6:   Let $\bar{\sigma}$ be the real number determined by the Cantor sequence σ of Lemma 4. Then $\bar{\sigma}$ is an upper bound for A.

proof: By Lemma 3, for all $x \in A$ and $n \in N_1$, $x < f(k_n/2^n)$. But $k_n/2^n = \sigma(n)$, so that for all $x \in A$ and $n \in N_1$,

$$x < f(\sigma(n)) = \overline{\sigma(n)}.$$

By Lemma 1, $\bar{\sigma} \geq x$ for all $x \in A$, hence $\bar{\sigma}$ is an upper bound for A, and Lemma 6 is proved.

lemma 7:   If u, v are real numbers and $u > v$, then there is a natural number k such that $u > v + f(1/2^k)$.

proof: Let $u = \bar{\rho}$, $v = \bar{\tau}$; since $u > v$, $\rho - \tau$ is a positive Cantor sequence. Let (m,r) be a test pair for $\rho - \tau$. There is a natural number k such that $1/2^k < r$ (why?). But then, $(m, r - 1/2^k)$ is a test pair for the sequence $\rho - \widehat{(1/2^k)} - \tau$. For, if $n \geq m$, then

$$\left(\rho - \left(\frac{\hat{1}}{2^k}\right) - \tau\right)(n) = \rho(n) - \left(\frac{\hat{1}}{2^k}\right)(n) - \tau(n)$$

$$= \rho(n) - \tau(n) - \frac{1}{2^k}.$$

But $\rho(n) - \tau(n) \geq r$; therefore $\rho(n) - \tau(n) - 1/2^k \geq r - 1/2^k$. Hence,

$\rho - \widehat{(1/2^k)} - \tau$ is positive, $\bar{\rho} - \overline{(1/2^k)} - \bar{\tau} > \bar{0}$, and $\bar{\rho} - \bar{\tau} > \overline{(1/2^k)} = f(1/2^k)$.

This completes the proof of Lemma 7.

lemma 8: $\bar{\sigma}$ (defined in Lemmas 5 and 6) is a supremum for A.

proof: By Lemma 6, $\bar{\sigma}$ is an upper bound for A; therefore it suffices to prove that if $\bar{\tau}$ is a real number and $\bar{\tau} < \bar{\sigma}$, then $\bar{\tau}$ is not an upper bound for A.

By Lemma 7 there is a natural number n such that

$$\bar{\tau} + f\left(\frac{1}{2^n}\right) < \bar{\sigma},$$

or,

$$\bar{\tau} < \bar{\sigma} - f\left(\frac{1}{2^n}\right).$$

By Lemma 5, for each $m \in N_1$,

$$\bar{\sigma} \le f\left(\frac{k_m}{2^m}\right);$$

hence, in particular, $\bar{\sigma} \le f(k_n/2^n)$. Then

$$\bar{\tau} < \bar{\sigma} - f\left(\frac{1}{2^n}\right) \le f\left(\frac{k_n}{2^n}\right) - f\left(\frac{1}{2^n}\right) = f\left(\frac{k_n - 1}{2^n}\right).$$

By Lemma 3, there is a $y \in A$ such that $y \ge f((k_n - 1)/2^n) > \bar{\tau}$, hence $\bar{\tau}$ is not an upper bound for A.

This completes the proof of Lemma 8 and Theorem 6.

q.e.d.

Definition 9. An ordered field F such that

> every nonempty subset of F which is bounded above has a supremum in F

is a *complete, ordered field.*

corollary 1: The field R is a complete, ordered field.

proof: Theorem 6 and Definition 9.

corollary 2: The ordered field R is not complete.

proof: Example 2, page 230.∗

EXERCISES

Prove:

1. That every nonempty subset of a complete, ordered field which is bounded below has an infimum.

2. If $a \in \mathbb{R}$ then there exist nonempty subsets A, B of R^* such that $a = \sup A = \inf B$.

3. Let A, B be nonempty subsets of R^* such that $R^* = A \cup B$ and for each $x \in A$ and for each $y \in B$, $x \leq y$. Prove that $\sup A = \inf B$.

4. Let S be a nonempty subset of a complete, ordered field and let $s = \sup S$. Prove: If $b < a$, then there is an element $s \in S$ such that $b < s \leq a$.

5.4.   ROOTS OF REAL NUMBERS

As a consequence of Exercise 2, page 204, it follows that $\mathbb{R}$ contains a square root of every positive rational number. But now, using the completeness of $\mathbb{R}$ we can deduce the stronger result:

* Theorem 7. (i) If a is a positive real number and if n is a positive integer, then there is one and only one positive real number b such that $b^n = a$.

(ii) If a is negative and n is an odd positive integer, there is one and only one negative real number b such that $b^n = a$. (In either case, b is the *n-th root of a*, and we write "$b = \sqrt[n]{a}$." For the case $n = 2$, $b = \sqrt{a}$.)

proof: If $n = 1$, the theorem is trivial. Therefore we assume that $n \geq 2$. Note that the proof of (ii) is an immediate consequence of (i). Hence, we restrict ourselves to a proof of the first part of the theorem.

The proof that there is at most one n-th root of a is easy. Indeed, suppose b, c are positive real numbers such that $b^n = c^n = a$. Then,

$$b^n - c^n = (b - c)(b^{n-1} + b^{n-2}c + \ldots + c^{n-1}).$$

Since b, c are positive, so is $b^{n-1} + b^{n-2}c + \ldots + c^{n-1}$, whence $b - c = 0$ and $b = c$.

The proof of the existence of an n-th root of a requires a little more effort. We begin by proving three lemmas.

lemma 1: For each real number $u > 1$, and for each integer $n \geq 2$, there is a $v \in \mathbb{R}$ such that
(i) $v^n < u$, and
(ii) $1 < v < u$.

proof of lemma 1: By induction. Since $u > 1$, $u - 1 > 0$ and $(u - 1)^2 = u^2 - 2u + 1 > 0$. Therefore,

$$4u < u^2 + 2u + 1 = (u + 1)^2,$$

whence

$$u < \left(\frac{u+1}{2}\right)^2,$$

and

$$\frac{1}{u} > \left(\frac{2}{u+1}\right)^2.$$

Consequently,

$$u = u^2 \cdot \frac{1}{u} > \left(\frac{2}{u+1}\right)^2 \cdot u^2 = \left(\frac{2u}{u+1}\right)^2.$$

Thus, setting $v = 2u/(u+1)$ we find that (i) holds for the case $n = 2$. To prove that (ii) holds for our choice of v, note that since $1 < u$ we have

$$1 + u < u + u = 2u$$

and

$$2 = 1 + 1 < 1 + u;$$

therefore,

$$2u < (1 + u)u,$$

so that

$$1 + u < 2u < (1 + u)u.$$

Hence,

$$1 < \frac{2u}{1 + u} < u.$$

Now assume the lemma holds for $k \geq 2$. Then, if u is a real number, $u > 1$, there is a $v \in \mathbb{R}$ such that

$$1 < v < u \text{ and } v^k < u.$$

But, again using the induction hypothesis, there is a $w \in \mathbb{R}$ such that

$$1 < w < v \text{ and } w^k < v.$$

Hence,

$$w^{k+1} < vw < v^2.$$

Since $k \geq 2$ and $v > 1$, then $v^2 \leq v^k$, so that

$$w^{k+1} < v^k < u$$

and $1 < w < u$. Therefore the lemma holds for all integers $n \geq 2$.

q.e.d. Lemma 1.

lemma 2: If a is a positive real number and n is a positive integer then the set $A = \{x \mid x \in \mathbb{R} \text{ and } x > 0 \text{ and } x^n < a\}$ is nonempty.

proof of lemma 2: If $a \geq 1$ then $1 \in A$. If $a < 1$ then

$$a^2 = a \cdot a < a \cdot 1 = a,$$

so $a \in A$. In either case, $A \neq \phi$.

q.e.d. Lemma 2.

lemma 3: The set A (of Lemma 2) is bounded above.

proof of lemma 3: We claim that $1 + a$ is an u.b. for A. For, suppose $y > 1 + a$; then $y^n > (1 + a)^n > a$, hence $y \notin A$. Consequently, if $z \in A$ then $z \leq 1 + a$ and $1 + a$ is an upper bound for A.

q.e.d. Lemma 3.

The proof of Theorem 7 is completed as follows:

Since $A \neq \phi$ is bounded above, by the completeness of $\mathbb{R}$, A has a supremum b, in $\mathbb{R}$. Since a and b^n are real numbers, one and only one of

$$b^n < a, \; b^n > a, \; b^n = a$$

holds. It suffices to rule out the first two alternatives to conclude the proof.

$b^n < a$ is false. For, if $b^n < a$, then $1 < a/b^n$ and by Lemma 1 there is a $v \in \mathbb{R}$ such that

$$v^n < \frac{a}{b^n} \text{ and } 1 < v < \frac{a}{b^n}.$$

From the first inequality we deduce $v^n b^n = (vb)^n < a$. Further, since v and b are both positive, so is vb, hence $vb \in A$. Since $b = \sup A$, $vb \leq b$. On the other hand, since $1 < v$, $b < vb$, a contradiction. Hence, $b^n < a$ is impossible.

Finally, $a < b^n$ is false. Indeed, if $a < b^n$, then $1 < b^n/a$ and again by Lemma 1 there is a $w \in \mathbf{R}$ such that

$$w^n < \frac{b^n}{a} \text{ and } 1 < w < \frac{b^n}{a}.$$

From the first inequality we have

$$a < \frac{b^n}{w^n} = \left(\frac{b}{w}\right)^n.$$

Now b/w is an upper bound for A. For, if $y > b/w$ then $y^n > (b/w)^n > a$; and this shows that if $x \in A$ then $x \leq b/w$. On the other hand, since $1 < w$, $b/w < b$. This means that an upper bound for A is less than $\sup A$, a contradiction. Hence, $a < b^n$ is false.

The only alternative is $b^n = a$. This concludes the proof of part (i) of Theorem 7.

EXERCISE

Prove part (ii) of Theorem 7.

Corollary: The field $\mathbf{R}$ is an ordered field with respect to one and only one subset.

proof: We know that $\mathbf{R}$ is an ordered field with respect to $\mathbf{R}^+$ (see Definition 21 and Theorem 21, Chapter 4). Suppose $\mathbf{R}$ is an ordered field with respect to a subset P. It will be proved that $P = \mathbf{R}^+$.

Let $u \in \mathbf{R}^+$; by Theorem 7 there is an $a \in \mathbf{R}$, $a \neq 0$, such that $u = a^2$. By Exercise 1, page 216, $a^2 \in P$, i.e., $u \in P$, hence $\mathbf{R}^+ \subset P$. On the other hand, if $u \notin \mathbf{R}^+$ and $u \neq 0$ then $-u \in \mathbf{R}^+$. Again, by Theorem 7, there is a $b \in \mathbf{R}$ such that $-u = b^2$. Then $b^2 \in P$ whence $-u \in P$ and $u \notin P$. Finally, if $u = 0$, then by definition of P, $u = 0 \notin P$. Therefore, $P \subset \mathbf{R}^+$ and $\mathbf{R}^+ = P$.

q.e.d.∗

EXAMPLE

Let S be the subset of R defined by

$$S = \{a + b\sqrt{2} \mid a, b \in R^*\}.$$

It has been proved (Exercise 1, page 227) that S is a subfield of R. We exhibit two subsets of S with respect to which S is an ordered field. The first is

$$P_1 = S \cap R^+.$$

(i) Let $a + b\sqrt{2} \in S$; then $a + b\sqrt{2} \in R$ and since R is an ordered field with respect to R^+, exactly one of

$$a + b\sqrt{2} \in R, \quad a + b\sqrt{2} = 0, \quad -(a + b\sqrt{2}) \in R$$

holds.   Hence, exactly one of

$$a + b\sqrt{2} \in P_1, \quad a + b\sqrt{2} = 0, \quad -(a + b\sqrt{2}) \in P_1$$

holds, and the first condition of Definition 2 is satisfied.

(ii) It is left to the reader to verify that if $a + b\sqrt{2}$ and $c + d\sqrt{2}$ are in P_1 then so are $(a + b\sqrt{2}) + (c + d\sqrt{2})$ and $(a + b\sqrt{2})(c + d\sqrt{2})$. Therefore, S is an ordered field with respect to P_1.   Next, let

$$P_2 = \{a - b\sqrt{2} \mid a + b\sqrt{2} \in P_1\}.$$

(i) For each $a - b\sqrt{2}$ in S we have

$$a - b\sqrt{2} \in P_2 \text{ if and only if } a + b\sqrt{2} \in P_1,$$
$$a - b\sqrt{2} = 0 \text{ if and only if } a + b\sqrt{2} = 0 \text{ (why?)},$$
$$-(a - b\sqrt{2}) \in P_2 \text{ if and only if } -(a + b\sqrt{2}) \in P_1,$$

hence condition (i) of Definition 2 holds.

(ii) Suppose $a - b\sqrt{2}$ and $c - d\sqrt{2}$ are in P_2.   Then $a + b\sqrt{2}$ and $c + d\sqrt{2}$ are in P_1, whence so is

$$(a + b\sqrt{2})(c + d\sqrt{2}) = (ac + 2bd) + (ad + bc)\sqrt{2}.$$

By definition of P_2,

$$(ac + 2bd) - (ad + bc)\sqrt{2} = (a - b\sqrt{2})(c - d\sqrt{2}) \in P_2.$$

The reader may verify that $(a - b\sqrt{2}) + (c - d\sqrt{2}) \in P_2$. Finally, $P_1 \neq P_2$ since $\sqrt{2} \in P_1$ and $\sqrt{2} \notin P_2$.

5.5. MORE THEOREMS ON ORDERED AND COMPLETE, ORDERED FIELDS

From the point of view of many branches of mathematics, the essential features of the real numbers are summed up in the statement:

R is a complete, ordered field.

Hence, for many purposes, any complete, ordered field will do as well as R. This leads to the conjecture that in some important respects all complete, ordered fields are indistinguishable. A proof of such a theorem will be given in Section 5.6; the purpose of the present section is to provide the tools for the proof. It is suggested that the reader review the definitions of upper bound, lower bound, supremum, infimum and complete, ordered field before proceeding.

Definition 10. Let A, B be subsets of a field K. Then

$$A + B = \{x + y \mid x \in A \text{ and } y \in B\}.$$

Theorem 8. Let A, B be subsets of an ordered field K. If $a = \sup A$ and $b = \sup B$, then $a + b = \sup (A + B)$.

proof: First, $a + b$ is an upper bound for $A + B$. For, if $u \in A + B$, then $u = x + y$ where $x \in A$ and $y \in B$. Since $a = \sup A$ and $b = \sup B$, it follows that $x \leq a$ and $y \leq b$. Hence, $u = x + y \leq a + b$, and $a + b$ is an upper bound for $A + B$.

Next, to show that $a + b = \sup (A + B)$ we prove:

If $z < a + b$, then z is not an upper bound for $A + B$.

Since $z - b < a$ and since $a = \sup A$, $z - b$ is not an upper bound for A. Hence, there is an $x \in A$ such that $z - b < x \leq a$. Therefore,

$$z < x + b \leq a + b.$$

Similarly, since $z - x < b$ and b is an upper bound for B, there is a $y \in B$ such that $z - x < y \leq b$. Consequently,

$$z < x + y \leq x + b \leq a + b.$$

Thus, $x + y$ is an element in $A + B$ such that $z < x + y$, hence z is not an upper bound for $A + B$.

q.e.d.

Definition 11. Let K be an ordered field and let A, B be nonempty subsets of K containing only nonnegative elements. Then

$$AB = \{x \cdot y \mid x \in A \text{ and } y \in B\}.$$

Theorem 9. Let K be an ordered field and let A, B be nonempty subsets of K containing only nonnegative elements. If $a = \sup A$ and $b = \sup B$, then $ab = \sup AB$.

proof: If a or b is zero, then obviously $0 = ab = \sup AB$ by the definition of AB.

Suppose $a \neq 0$ and $b \neq 0$; then a and b are both positive. We prove, first, that ab is an upper bound for AB. For, let $u \in AB$. Then there are elements $x \in A$ and $y \in B$ such that $u = xy$. But, since $a = \sup A$ and $b = \sup B$, $x \leq a$ and $y \leq b$. Further, since x, y, a and b are all nonnegative, we deduce $u = xy \leq ab$. Hence, ab is an upper bound for AB.

Now let $z < ab$. Since b is positive, $zb^{-1} < a$. Since $a = \sup A$, there is an $x \in A$ such that $zb^{-1} < x \leq a$. Then $z < xb \leq ab$. Similarly, since x is positive, $zx^{-1} < b$. And, since $b = \sup B$, there is a $y \in B$ such that $zx^{-1} < y \leq b$. Therefore,

$$z < xy \leq xb \leq ab.$$

Hence, z is not an upper bound for AB; therefore $ab = \sup AB$.

q.e.d.*

It will be convenient to have the following notation:

* Definition 12. Let S and T be sets and let η be a mapping of S into T. Then

$$\eta(S) = \{\eta(x) \mid x \in S\}.$$

lemma: Let F and G be ordered fields and let σ be an order-preserving isomorphism between F and G. If A is a subset of F and $u \in F$ is an upper bound for A, then $\sigma(u) \in G$ is an upper bound for $\sigma(A)$.

proof: Exercise.

Theorem 10. Let F and G be fields, and let σ be an isomorphism between F and G. If F is a complete, ordered field, so is G. In other words:

An isomorphic image of a complete, ordered field is a complete, ordered field.

proof: It has already been shown that G is an ordered field and that σ is an order-preserving isomorphism (Corollary to Theorem 3). All that is left is to prove G is complete.

To simplify the notation, we use the same symbol "$<$" to denote "less than" in both F and G.

Suppose $B \neq \phi$ is a subset of G, and B is bounded above. Let A be the subset of F such that $\sigma(A) = B$. Then A is also bounded above (why?) and, since F is complete, there is an $\bar{a} \in F$ such that $\bar{a} = \sup A$. We claim that $\sigma(\bar{a}) = \sup B$.

By the lemma, $\sigma(\bar{a})$ is an upper bound for $\sigma(A) = B$. It suffices to show that if $v < \sigma(\bar{a})$ then v is not an upper bound for B. For, if $v < \sigma(\bar{a})$ then since σ^{-1} is also order-preserving, $\sigma^{-1}(v) < \sigma^{-1}(\sigma(\bar{a})) = \bar{a}$. Since $\bar{a} = \sup A$, there is a $y \in A$ such that $\sigma^{-1}(v) < y \leq \bar{a}$. Then $v = \sigma(\sigma^{-1}(v)) < \sigma(y) \leq \sigma(\bar{a})$. And, since $\sigma(y) \in B$, it follows that v is not an u.b. for B. Therefore, $\sigma(\bar{a}) = \sup B$.

q.e.d.∗

Let K be a field, n a nonnegative integer. If $K = R$ or if $K = R$ and if $x \in K$, then $n \cdot x$ is a product of field elements, provided the word "integer" is interpreted in the ways discussed on pages 226 and 227. However, it may happen that K is neither R nor R and does not even contain these fields as subsets. In such a case, it may be impossible to regard any elements in K as integers. Nevertheless, we still want to assign a meaning to the expression "$n \cdot x$." The idea that we have in mind is the simple one of defining $n \cdot x$ in such a way that, say, for each $x \in K$,

$$3 \cdot x = x + x + x,$$
$$4 \cdot x = x + x + x + x,$$

"and so on." For this purpose, we require the following definition by recursion:

∗ Definition 13. Let K be a field, and for each $x \in K$ set

$$0 \cdot x = 0_K$$

where 0 is the integer, zero, and 0_K is the zero element of K. If m is any nonnegative integer, assume $m \cdot x$ has been defined and set

$$(m + 1) \cdot x = m \cdot x + x.$$

Then $n \cdot x$ is defined for all nonnegative integers n and for all $x \in K$.

To avoid ambiguity in the next two theorems, we shall let e be the unity element in K. If we were to use the customary symbol "1" to denote the unity element of K, the expression "$1 \cdot 1$" could be interpreted in two ways: namely, as a product of integers, and also in accordance with Definition 13. With e as the unity element, "$1 \cdot e$" has a meaning only in terms of Definition 13, and $1 \cdot 1$ is a product of integers. Observe that if $K = \mathcal{R}$ or $K = R$ and $x \in K$, then the meaning of "$n \cdot x$" in accordance with Definition 13 coincides with $n \cdot x$ as a product of elements in K.$_*$

In Exercise 2, page 172, Chapter 3, it was proved that the Archimedean Law holds for the rational numbers. The proof of the corresponding law for a complete, ordered field is slightly more subtle.

* Theorem 11. (The Archimedean Law for a complete, ordered field). If a, b are positive elements in a complete, ordered field F, then there is a positive integer n such that $n \cdot a > b$.

proof: By contradiction. Suppose $m \cdot a \leq b$ for all positive integers m. Then the set

$$A = \{m \cdot a \mid m \in Z \text{ and } m > 0\}$$

is a nonempty subset of F bounded above by b. Since F is complete, A has a supremum $u \in F$, and so $u \geq m \cdot a$ for all positive integers m. Consequently, $u \geq (k + 1) \cdot a$ for all positive integers k, and so

$$u \geq k \cdot a + 1 \cdot a = k \cdot a + a,$$

or

$$u - a \geq k \cdot a,$$

for all positive integers k. Thus,

$$k \cdot a \leq u - a < u,$$

and this is a contradiction, since $u = \sup A$. Therefore, there is a positive integer n such that $n \cdot a > b$.

q.e.d.$_*$

EXERCISES

Prove:

1. If m is a positive integer and x is a positive element in the ordered field K, then $m \cdot x$ is positive.

2. If m is a nonnegative integer and e is the unity element in the field K, and $x \in K$, then $m \cdot x = (m \cdot e) \cdot x$.

3. If a, b are positive elements in a complete, ordered field F, there exists a largest nonnegative integer m such that $m \cdot a < b$.

* Theorem 12. Let F be a complete, ordered field and let R' be the subfield of F, which is order-isomorphic with R. If a, $b \in F$ and $a < b$, then there is an $r \in R'$ such that $a < r < b$.

proof: Assume, for the moment, we have proved the following special case of the theorem:

(5.7) If $c \in F$, $0 < c$, then there is an element $t \in R'$ such that

$$0 < t < c.$$

First, suppose $0 < a$ Since $a < b$, $0 < b - a$ and by (5.7) there is an $s \in R'$ such that $0 < s < b - a$. By Exercise 3, above, there is a largest positive integer m such that $m \cdot s < b$. Since $s \in R'$, $m \cdot s \in R'$ and we need only show that $a < m \cdot s$.

By the choice of m,

(5.8) $$(m + 1) \cdot s = m \cdot s + s \geq v$$

Now suppose $m \cdot s \leq a$.   Then

$$(m + 1) \cdot s = m \cdot s + s \leq a + s < a + (b - a) - v,$$

in contradiction to (5.8). Therefore $a < m \cdot s < b$ where $m \cdot s \in R'$.

Next let $a < 0 < b$. In this case, obviously $r = 0$ does the trick.

Finally, let $a < b < 0$. Then $0 < -b < -a$, and by the first case there is an element $-r \in R'$ such that $-b < -r < -a$. Then $a < r < b$ where $r \in R'$.

To complete the proof, we verify the special case (5.7).

Let e be the unity element of F. Then $e \in R'$, e is positive, and so for that matter, is $n \cdot e$ for all positive integers n. Moreover, $(n \cdot e)^{-1} \in R'$, and $(n \cdot e)^{-1}$ is positive for each integer $n > 0$. Since $c > 0$, by the Archimedean Law there is a positive integer m such that

$$m \cdot c > e,$$

whence, by Exercise 2, above.

$$(m \cdot e) \cdot c > e.$$

Therefore

$$c > e \cdot (m \cdot e)^{-1} > 0,$$

where $e \cdot (m \cdot e)^{-1} \in R'$.

q.e.d.*

From now on, the unity element in all fields under consideration will be denoted by "1".

5.6.   THE ISOMORPHISM OF COMPLETE, ORDERED FIELDS

The purpose of this section is to prove the following

* Theorem 13.   Let F_1 and F_2 be complete, ordered fields and let R_1' and R_2' be the subfields of F_1 and F_2, respectively, which (by Corollary 3 to Theorem 5) are order-isomorphic to each other and to R. Further, let σ be an order-preserving isomorphism between R_1' and R_2'. Then there exists an order-preserving isomorphism σ^* between F_1 and F_2, which is an extension of σ. (See Definition 24, Chapter 1.)

For convenience, we denote the binary operations of both F_1 and F_2 by "+", "·" (or, juxtaposition); also, we use "$<$" as the symbol for "less than" in both fields.

The following definition has an important role in the proof.

Definition 14.   Let F be a complete, ordered field, and let R' be the subfield of F which is order-isomorphic with R. Then, for each nonnegative element $a \in F$,

$$L_a = \{x \mid x \in F \text{ and } 0 \leq x \leq a\} \cap R'.$$

Note: $a = \sup L_a$ (why?).

The crucial step in the proof of Theorem 13 is the following

lemma: For each $a \in F_1$, $a \geq 0$, there is an $\bar{a} \in F_2$ such that

$$\sigma(L_a) = L_{\bar{a}}.$$

proof: Since L_a is bounded above (Definition 6) and nonempty and since σ is an order-preserving isomorphism between R_1' and R_2', $\sigma(L_a) \neq \phi$ and is also bounded above. Since F_2 is complete, there is an $\bar{a} \in F_2$ such that $\bar{a} = \sup \sigma(L_a)$. Moreover, since the elements of L_a are all non-negative, so are the elements of $\sigma(L_a)$, hence $\bar{a} \geq \bar{0}$. We claim that

$$\sigma(L_a) = L_{\bar{a}}.$$

To prove that $\sigma(L_a) \subset L_{\bar{a}}$ let $\bar{x} \in \sigma(L_a)$. Since $\bar{a} = \sup \sigma(L_a)$ it follows that $\bar{x} \leq \bar{a}$. Further, there is a unique $x \in L_a$ such that $\sigma(x) = \bar{x}$. But then $x \in R_1'$ and $x \geq 0$; and since σ is an order-preserving isomorphism between R_1' and R_2' it follows that $\sigma(x) = \bar{x} \in R_2'$ and $\bar{x} \geq \bar{0}$. Thus,

$$\bar{0} \leq \bar{x} \leq \bar{a} \text{ and } \bar{x} \in R_2',$$

whence $\bar{x} \in L_{\bar{a}}$. Therefore, $\sigma(L_a) \subset L_{\bar{a}}$.

To prove the reverse inclusion, let $\bar{y} \in L_{\bar{a}}$. Then $\bar{y} \in R_2'$ and $\bar{0} \leq \bar{y} \leq \bar{a}$; hence there is a unique $y \in R_1'$ such that $\sigma(y) = \bar{y}$ and $y \geq 0$. If we can show that $y \leq a$, it follows that $y \in L_a$, whence $\sigma(y) = \bar{y} \in \sigma(L_a)$, therefore $L_{\bar{a}} \subset \sigma(L_a)$.

Suppose $y > a$. By Theorem 12, there is an $r \in R_1'$ such that $y > r > a$. Since $a = \sup L_a$, r is an upper bound for L_a. Therefore $\sigma(r)$ is an upper bound for $\sigma(L_a)$ (lemma, page 245). But $\bar{a} = \sup \sigma(L_a)$, hence $\sigma(r) \geq \bar{a}$. Finally, since $y > r$ and since σ is order-preserving, we deduce $\sigma(y) = \bar{y} > \sigma(r) \geq \bar{a}$, which contradicts the fact that $\bar{a} \geq \bar{y}$. Therefore, $y \leq a$ and, as remarked above, we have $L_{\bar{a}} \subset \sigma(L_a)$.

q.e.d. Lemma.

proof of theorem 13: The proof is carried out by constructing the desired mapping σ^* in several steps and then checking that σ^* is an isomorphism.

For each $a \in F_1$, $a \geq 0$, set

$$\sigma^*(a) = \bar{a} \text{ where } \bar{a} = \sup \sigma(L_a).$$

Since $a = b$ if and only if $L_a = L_b$, since $\sigma : R_1' \longrightarrow R_2'$ is an isomorphism and since each $\sigma(L_a)$ has exactly one supremum in F_2, σ^* is one-one. On the other hand, since σ^{-1} is also an order-preserving isomorphism, for each $\bar{a} \in F_2$, $\bar{a} \geq \bar{0}$, there is an $a \in F_1$, $a \geq 0$, such that

$$a = \sup \sigma^{-1}(L_{\bar{a}}),$$

and by the lemma, $\sigma^{-1}(L_{\bar{a}}) = L_a$. Hence, σ^* is onto. Moreover, if $a \in R_1'$, $a \geq 0$, we claim that

$$\sigma^*(a) = \sigma(a),$$

i.e., σ^* is an extension of σ. For, if $a \in R_1'$ and $a \geq 0$, then $a \in L_a$, whence $\sigma(a) \in \sigma(L_a) = L_{\bar{a}}$. Further, $\bar{a} = \sup L_{\bar{a}}$ and so $\sigma(a) \leq \bar{a}$. If $\sigma(a) < \bar{a}$, then there is an $\bar{r} \in R_2'$ such that $\sigma(a) < \bar{r} < \bar{a}$ (Theorem 12), whence $\bar{r} \in L_{\bar{a}} = \sigma(L_a)$. Hence, there is an $r \in L_a$ such that $\sigma(r) = \bar{r}$; and since σ is order-preserving, $a < r$. This is a contradiction, since $a = \sup L_a$ and $r \in L_a$.

The definition of σ^* is completed by setting

$$\sigma^*(a) = -\sigma^*(-a)$$

for all negative $a \in F_1$. Clearly, $\sigma^*(x)$ is positive if and only if x is positive, hence σ^* is order-preserving; and $\sigma^*(x) = \sigma(x)$ for all $x \in R_1'$.
Next,

(5.9) $\sigma^*(a + b) = \sigma^*(a) + \sigma^*(b), \quad a, b \in F_1.$

case 1: $a \geq 0$ and $b \geq 0$. The result, equation (5.9), follows at once by the lemma and by Theorem 8.

case 2: $a < 0$ and $b < 0$. Then $-a > 0$, $-b > 0$, $-(a + b) > 0$, hence

$$\begin{aligned}
\sigma^*(a + b) &= -\sigma^*(-(a + b)) \\
&= -(\sigma^*(-a) + \sigma^*(-b)) \\
&= -\sigma^*(-a) - \sigma^*(-b) \\
&= \sigma^*(a) + \sigma^*(b).
\end{aligned}$$

case 3: $a < 0$, $b \geq 0$ and $a + b > 0$. Set $a + b = p$; then

$$\begin{aligned}
\sigma^*(b) &= \sigma^*(p + (-a)) \\
&= \sigma^*(p) + \sigma^*(-a),
\end{aligned}$$

whence

$$\begin{aligned}
\sigma^*(a + b) = \sigma^*(p) &= -\sigma^*(-a) + \sigma^*(b) \\
&= \sigma^*(a) + \sigma^*(b).
\end{aligned}$$

case 4: $a < 0$, $b \geq 0$ and $a + b < 0$. Then $-b \leq 0$, $-a > 0$ and $(-a) + (-b) > 0$. Now proceed as in Case 3.
Finally,

(5.10) $\sigma^*(ab) = \sigma^*(a)\sigma^*(b), \quad a, b \in F_1.$

case 1: $a \geq 0$ and $b \geq 0$. Equation (5.10) follows at once from the lemma and Theorem 9.

case 2: $a < 0$ and $b < 0$. Then $ab = (-a)(-b) > 0$ and

$$
\begin{aligned}
\sigma^*(ab) &= \sigma^*((-a)(-b)) \\
&= \sigma^*(-a)\sigma^*(-b) \\
&= (-\sigma^*(-a))(-\sigma^*(-b)) \\
&= \sigma^*(a)\sigma^*(b).
\end{aligned}
$$

case 3: $a < 0$ and $b \geq 0$. Then $(-a)b = -(ab) \geq 0$ and

$$
\begin{aligned}
\sigma^*(ab) &= -\sigma^*(-(ab)) \\
&= -\sigma^*((-a)b) \\
&= -(\sigma^*(-a)\sigma^*(b)) \\
&= (-\sigma^*(-a))\sigma^*(b) \\
&= \sigma^*(a)\sigma^*(b).
\end{aligned}
$$

Thus, in every case, equation (5.10) holds and the theorem is proved.

$$\text{q.e.d.}_*$$

The distressing thought may have occurred to the reader, "What if there is only one complete, ordered field, namely the field $\mathbb{R}$ so laboriously constructed in this text?" If this were the case, then Theorem 13 would be established at once by observing that the mapping $g: \mathbb{R} \longrightarrow \mathbb{R}$ defined by

$$g(x) = x, \quad x \in \mathbb{R},$$

does the trick. There are, however, several different ways of constructing "the" real numbers. For instance, beginning with a set $\overline{N}$ order-isomorphic with the natural numbers N, one constructs a set of unsigned rationals $\overline{R}$. $\overline{R}$ can be ordered and its elements can be added, multiplied and divided (except for division by 0). Subtraction, $a - b$, is restricted to those cases in which $a > b$ in the order relation $>$ on $\overline{R}$. From $\overline{R}$ one then constructs the unsigned reals $\overline{\mathbb{R}}$ by the use of Cantor sequences of elements in $\overline{R}$. Finally, from $\overline{\mathbb{R}}$ one constructs a complete field $\mathbb{R}'$ by means of equivalence classes of ordered pairs of elements in $\overline{\mathbb{R}}$. By Theorem 13, $\mathbb{R}$ and $\mathbb{R}'$ are order-isomorphic.

Still another method for constructing the real numbers is the method of *Dedekind Cuts* in the rationals R. The reader will find the method discussed in *A Survey of Modern Algebra* (Revised Edition) by Birkhoff and MacLane.

5.7.   THE COMPLEX NUMBERS

*Let C be the set $R \times R$. We define a pair of mappings, $\oplus$ and $\odot$, of $C \times C$ into C, by:

$$(a,b) \oplus (c,d) = (a + c, b + d), \qquad (a,b), (c,d) \in C,$$

and

$$(a,b) \odot (c,d) = (ac - bd, ad + bc), \qquad (a,b), (c,d) \in C,$$

respectively. From their definitions, it is clear that $\oplus$ and $\odot$ are mappings of $C \times C$ into C; therefore, they are binary operations on C.

Definition 15. The elements in C are *complex numbers*.

Theorem 14. With the binary operations $\oplus$ and $\odot$, C is a field whose zero element is $(0,0)$ and whose unity element is $(1,0)$.

proof: Clearly, $\oplus$ is commutative and associative. Since $(a,b) \oplus (0,0) = (a + 0, b + 0) = (a,b)$, $(0,0)$ is the zero element of C. And, since $(a,b) \oplus (-a,-b) = (a - a, b - b) = (0,0)$, the inverse of (a,b) with respect to $\oplus$ is $(-a,-b)$. We write: "$-(a,b) = (-a,-b)$."

The reader can verify easily that $\odot$ is also commutative and associative. Further,

$$(a,b) \odot (1,0) = (a \cdot 1 + b \cdot 0, a \cdot 0 + b \cdot 1) = (a,b),$$

so that $(1,0)$ is the unity element.

The distributive law holds in C:

$$
\begin{aligned}
(a,b) \odot ((c,d) \uplus (e,f)) &= (a,b) \odot (c + e, d + f) \\
&= (a(c + e) - b(d + f), a(d + f) + b(c + e)) \\
&= (ac - bd, ad + bc) \oplus (ae - bf, af + be) \\
&= ((a,b) \odot (c,d)) \oplus ((a,b) \odot (e,f)).
\end{aligned}
$$

Finally, if $(a,b) \neq (0,0)$ then (a,b) has an inverse with respect to $\odot$ in C. For, $\left(\dfrac{a}{a^2 + b^2}, \dfrac{-b}{a^2 + b^2} \right) \in C$ and

$$(a,b) \odot \left(\frac{a}{a^2 + b^2}, \frac{-b}{a^2 + b^2} \right) = (1,0).$$

q.e.d.*

In the field of complex numbers we can prove the following generalization of Exercise 2, page 204:

*Theorem 15. Every element $(a,b) \in C$, $(a,b) \neq (0,0)$ has two square roots in C. In other words, if $(a,b) \neq (0,0)$, then there exist (x,y), $(u,v) \in C$, $(x,y) \neq (u,v)$, such that

$$(x,y)^2 = (x,y) \odot (x,y) = (a,b),$$

and

$$(u,v)^2 = (u,v) \odot (u,v) = (a,b).$$

proof: It is easy to verify that the ordered pairs

$$\left(\sqrt{\frac{a + \sqrt{a^2 + b^2}}{2}}, \; \frac{b}{2\sqrt{\dfrac{a + \sqrt{a^2 + b^2}}{2}}} \right),$$

and

$$\left(-\sqrt{\frac{a + \sqrt{a^2 + b^2}}{2}}, \; \frac{-b}{2\sqrt{\dfrac{a + \sqrt{a^2 + b^2}}{2}}} \right)$$

are in C and that their squares are (a,b).

ч.e.d.⁂

EXERCISE

How were the two complex numbers of Theorem 15 obtained?

* Unlike $\mathbb{R}$, C cannot be made into an ordered field. For, if it were possible to do so, there would be a nonempty subset P (the set of positive elements) of C such that

(i) for all $(a,b) \in C$, $(a,b) \neq (0,0)$, exactly one of

$$(a,b) \in P \text{ or } -(a,b) \in P$$

holds; and

(ii) if (a,b), $(c,d) \in P$ then $(a,b) \oplus (c,d)$ and $(a,b) \odot (c,d) \in P$.

Now, since $(1,0)$ is the unity element of C, $(1,0) \in P$, hence $(-1,0) = -(1,0) \notin P$.

Consider the element $(0,1)$. By (i), exactly one of

$$(0,1) \in P \text{ or } -(0,1) = (0,-1) \in P$$

must hold.   If $(0,1) \in P$ then by (ii)

$$(0,1) \odot (0,1) = (-1,0) \in P,$$

a contradiction.   On the other hand, if $(0,-1) \in P$, then by (ii)

$$(0,-1) \odot (0,-1) = (-1,0) \in P,$$

the same contradiction.   Thus the assumption that C can be made into an ordered field always yields a contradiction.∗

Although C cannot be ordered as a field, it does contain a subfield isomorphic with R.

Theorem 16.   Let $R^* = \{(a,b) \mid (a,b) \in C \text{ and } b = 0\}$.   Then R and R^* are isomorphic and, furthermore, R^* is a subfield of C.

proof: Define a mapping $g : R \longrightarrow C$ by

$$g(a) = (a,0), \quad a \in R.$$

Since $(a,0) = (b,0)$ if and only if $a = b$, g is a one-one mapping of R into C and onto R^*.   Further,

$$g(a + b) = (a + b, 0) = (a,0) \oplus (b,0) = g(a) \oplus g(b)$$

and

$$g(ab) = (ab,0) = (a,0) \odot (b,0) = g(a) \odot g(b),$$

and so g is an isomorphism.   Therefore (Theorem 1) R^* is a field.

To prove that R^* is a subfield of C, note first $(1,0) \in R^*$ and $(1,0) \neq (0,0)$.   If $(x,0), (y,0) \in R^*$ then $(x,0) \oplus (-(y,0)) = (x - y, 0) \in R^*$.   And if $(x,0) \neq (0,0)$ then $(1/x,0) \in R^*$.   Thus the conditions of Theorem 4 are satisfied and R^* is a subfield of C.

$$\text{q.e.d.}∗$$

Note: R^* is a complete, ordered field different from (but, of course, isomorphic to) R.

We conclude this brief discussion of the complex numbers by introducing the customary notation.

If (a,b) is a complex number, then $(a,b) = (a,0) \oplus (0,b)$, and $(0,b) = (b,0) \odot (0,1)$.   Consequently,

$$(a,b) = (a,0) \oplus ((b,0) \odot (0,1)).$$

For each "$(x,0)$" we shall henceforth write "x", and in place of "$(0,1)$"

we write "i". Also we replace "$\oplus$" and "$\odot$" by "$+$" and "$\cdot$", respectively. Then

$$(a,b) = a + bi.$$

Note that

$$i^2 = i \cdot i = (0,1) \odot (0,1) = (-1,0) = -1;$$

further

$$\begin{aligned}(a + bi) + (c + di) &= (a,b) \oplus (c,d) \\ &= (a + c, b + d) \\ &= (a + c) + (b + d)i,\end{aligned}$$

and

$$\begin{aligned}(a + bi) \cdot (c + di) &= (a,b) \odot (c,d) \\ &= (ac - bd, ad + bc) \\ &= (ac - bd) + (ad + bc)i.\end{aligned}$$

Thus we may compute with expressions "$a + bi$", "$c + di$", etc., as in elementary algebra.

EXERCISE

What, if anything, is wrong with

$$1 = \sqrt{1} = \sqrt{(-1)(-1)} = \sqrt{-1}\,\sqrt{-1} = -1?$$

INDEX

Absolute value function:
  for integers, 148–149
    definition, 148
  for rational numbers, definition, 172
Addition: (*see also* Arithmetic)
  of integers, 138–139, 148
  of natural numbers, 106–110, 112–115, 127–128
    table, 109
  of rational numbers, 163
  of real numbers, 204
Addition function:
  for integers, 147–148
    definition, 147
  for natural numbers, 125–127
    definition, 125
Additive inverse:
  of integers, 146
    definition, 142

Additive inverse (*Contd.*)
  of rational numbers, definition, 165
  of sequence of rational numbers, definition, 181
Algebra of sets, 25–34
"And," in logic, definition, 20
Antecedent, 23–24
Archimedean Law:
  for complete ordered field, 247
  for rational numbers, 172
Arithmetic: (*see also* Addition; Division; Multiplication; Subtraction)
  of Cantor sequence, 193–194
  of null sequence, 196–199
  of rational numbers, 161–172
  of real numbers, 199–200, 204
  of sequence of rational numbers, 180–187

Associative law:
  for addition:
    of integers, 138–139, 148
    of natural numbers, 107–108, 127–128
    of rational numbers, 163
    of real numbers, 204
  for multiplication:
    of integers, 138–139, 148
    of natural numbers, 110
    of rational numbers, 163
Associativity of composition of mappings, 58
Axiom of Extensionality, 14–15

Base, 122
Binary operation, 61–63
  definition, 62
  symbols, 226–227
Binary system for natural numbers, 85
Bound:
  for sequence of rational numbers, definition, 184
  for subset of ordered field, 228–238
Bound variable, 9
Bounded sequence:
  Cantor sequence, 192
  of rational numbers, 184–187

Cancellation law:
  for addition:
    of integers, 138–139
    of natural numbers, 113
    of rational numbers, 163
  for multiplication:
    of integers, 138–139
    of natural numbers, 114
    of rational numbers, 163

Cantor sequence of rational numbers, 187–194, 195, 198, 232–238 (*see also* Null sequence; Real numbers)
  definition, 189
  ordering, 205–209
Cardinality, definition, 101
Cartesian products, 46–47
Closeness of terms in a sequence, 179 (*see also* Cantor sequence)
Common divisor, definition, 151
Commutative law:
  for addition:
    of integers, 138–139, 148
    of natural numbers, 107
    of rational numbers, 163
    of real numbers, 204
  for multiplication:
    of integers, 138–139, 148
    of natural numbers, 110
    of rational numbers, 163
Commutativity of intersection, 26
Complement of sets, 29–34, 37
  definition, 30, 37
Complete ordered field, 245
  definition, 238
  isomorphism, 249–252
Completeness of real numbers, 227–239
Complex numbers, 253–256
  definition, 253
  notation, 255–256
Components of ordered pairs, 44–45
Composite of functions, 56–58
  definition, 56
Concept of set, 1–4
Conjunction, 20
Consequent, 23–24
Constant, 4–6
  definition, 4

Constant sequence of rational numbers:
  as Cantor sequence, 192
  definition, 182
Convergence test for sequence of rational numbers, 189–191
  definition, 189
Coprime integers, definition, 151
Copy (*see* Isomorphism)
Count, definition, 99
Counting, 98–101

Decimal system for natural numbers, 85
Denominator, 157
Difference:
  of integers, definition, 142
  of rational numbers, definition, 165
  of sequences of rational numbers, definition, 181
Disjunction, definition, 19
Distance between two points, 179
  (*see also* Measurement)
Distributive laws:
  in algebra of sets, 28–29
  for multiplication:
    of integers, 138–139
    of natural numbers, 110–111
    of rational numbers, 163
Division:
  of integers, 149–154
  of natural numbers, 132
  of rational numbers, 165–167
Division algorithm theory of Euclid, 150
Divisor, 149
Domain of a relation, 47

Element of a set:
  definition, 10

Element of a set (*Contd.*)
  final, 89–92
  leading, 80
Elements of the theory of sets, 1–73
Empty set, 16–18, 75, 76
  definition, 17
Equality, 9–10
  of sets, 14–15
Equals, definition, 9
Equivalence, 21–22
Equivalence class, definition, 72
Equivalence relation, 63–72
  definition, 63
Essential bound of sequence, 186–187
  definition, 186
Even integer, definition, 178
Existential quantifier, 8
Exponent, 122
Extensionality, Axiom, 14–15
Extension of a function, 55

Factor, 149
Field, 174–176 (*see also* Ordered field)
  definition, 174
  of rational numbers, relation to ordered field, 221–227
Final component of ordered pair, 44–45
Final element, 89–92
Finite Induction, 86–88 (*see also* Principle of Finite Induction)
Finite set, 101–106
  definition, 101
Fraction, definition, 157
Free variable, 9
Function, 49–63
  definition, 50, 52
Fundamental Theorem of Arithmetic, 155–156

Generalized associative law for addition and multiplication of natural numbers, 128

Generalized operations of integers, 147–157

Generalized product, 123–124

Generalized product function of integers, 148

Generalized sum, 125–127

Generalized sum function of integers, 148

"Greater than" relationship, 172 (*see also* Ordering)
  for integers, 140
  for natural numbers, 94
  for rational numbers, 167
  for real numbers, 210

Greatest common divisor, 151–153
  definition, 151

Hereditary set, definition, 78

Identity function, 51

Identity mapping, 51

"If and only if," 93–94

Image, 53

Implication, 23–24

"Implies," in logic, definition, 23

Improper subset, definition, 18

Indirect proof, 17

Induction, finite, 86–88 (*see also* Principle of Finite Induction)

Inequality, 94

Infimum, definition, 231

Infinite sequence, 117–119 (*see also* Sequence)
  definition, 117

Infinite set, definition, 101

Initial component of ordered pair, 44–45

Integers, 133–176
  addition, 137–139, 148
  definition, 134
  division, 149–153
  generalized operations, 147–157
  multiplication, 137–139, 148
  notation, 146
  ordering, 140–143
  properties, 133–147
  subtraction, 142

Integral domain, 173–176
  definition, 173

Intersection of sets, 25–27, 37, 42
  definition, 25, 26, 37

Interval, definition, 96

Inverse (*see* Additive inverse)

Inverse image, 53

Inverse relation, definition, 49

Isomorphic image, 215, 220, 246

Isomorphism, 143–144, 169–172, 214, 219 (*see also* Order-preserving isomorphism)
  of complete ordered fields, 249–252
  definition, 143
  for fields, 214–216

Later terms in a sequence, 179 (*see also* Cantor sequence)

Leading element, definition, 80

Least common multiple, 157

Left component of ordered pair, 45

"Less than" relationship, 172 (*see also* Ordering)
  for integers, 140
  for natural numbers, 94
  in ordered field, 218
  for rational numbers, 167
  for real numbers, 210

Like sequences, 195, 198

Logic, 18–24
Lower bound for subset of ordered field, 231

Mapping (*see* Function)
Mapping into, 50–52
 definition, 50
Mapping onto, definition, 52
Mathematical systems, 72–73 (*see also* Complex numbers; Integers; Natural numbers; Rational numbers; Real numbers)
Maximum, definition, 186
Measurement of distance, 179, 181–182, 194–195 (*see also* Sequence)
Modus Ponens, 24
Multiple, 149
Multiplication: (*see also* Arithmetic)
 of integers, 137–139, 148
 of natural numbers, 110–115
  table, 111
 of rational numbers, 163
Multiplication functions, 121–124, 127
 definition, 121
 of integers, 148
  definition, 147
Multiplicative inverse:
 of rational numbers, definition, 166
 of real numbers, definition, 212
Multiplicities of the real zeros, 129–130
Multiplicity, 157

Natural numbers, 74–131
 addition, 106–110, 112–115, 127–128

Natural numbers (*Contd.*)
 addition function, 125–127
 counting, 98–101
 definition, 81
 finite sets, 101–106
 multiplication, 110–115
 ordering, 89–98, 112–115
 Principle of Finite Induction, 115–117
 properties, 74–88
 relations of order, addition, and multiplication, 112–115
 sequence, 117–119
Negation, 21–22
Negative Cantor sequence, definition, 207
Negative elements of ordered field, definition, 216
Negative integer, definition, 141
Negative of a number (*see* Additive inverse)
Negative rational numbers, 168
 definition, 167
Negative real numbers, definition, 210
"Not," in logic, definition, 21
Notation:
 of complex numbers, 255–256
 of integers, 146
 of numbers and binary operations, 226–227
 of rational numbers, 160, 171–172
 of real numbers, 199, 226
 of sets, 10–12, 34–42
n-th term of a sequence, definition, 118
Null sequence, 194–199
 definition, 195
Null set, 16 (*see also* Empty set)
Nullity test for a sequence, definition, 195

Number of elements, definition, 101
Numbers: (*see also* Complex numbers; Integers; Natural numbers; Rational numbers; Real numbers)
    symbols, 226
Numerator, 157

Odd integer, definition, 178
One-one correspondence, 58–61
    definition, 59
One-one function, 58–61
One-one mapping, 58–61
"Or," in logic, definition, 19
Ordered fields, 214–220, 242–243
    complete, 245
        definition, 238
    definition, 216
    relation to field of rational numbers, 221–227
Ordered pairs, 43–45
    definition, 44
Ordering:
    of Cantor sequence, 205–209
    of integers, 140–143
    of natural numbers, 89–98, 112–115
    of rational numbers, 167–169
    of real numbers, 209–211
Order-preserving isomorphism, 222–226 (*see also* Isomorphism)
    of ordered field, 219–220
        definition, 219

Partition, 68–72
    definition, 68
PFI, 128–131 (*see also* Principle of Finite Induction)
Pigeon-hole principle, 105

Positive Cantor sequence, 205–209
    definition, 205
Positive elements of ordered field, definition, 216
Positive integer, definition, 141
Positive rational numbers, 167–168
    definition, 167
Positive real numbers, definition, 209
Power function, 122
Power set, 40–42
    definition, 40
Predicate in a symbol, 36–37
Prime integer, definition, 154
Principle of Finite Induction (PFI), 86–88, 115–117, 128–131
    for nonnegative integers, 144–145
Product:
    definition, 110
    of integers, definition, 137
    of rational numbers, definition, 160
    of real numbers, definition, 200
    of sequences of rational numbers, definition, 181
Proper subset, definition, 18

Quantifier, 7–9
Quotient, 150
    of rational numbers, definition, 166
Quotient field, 175–176

Range:
    of a relation, 47
    of a variable, 7
Rational numbers, 133, 157–176
    addition, 163
    arithmetic, 161–172
    definition, 158

Rational numbers (*Contd.*)
  division, 165–167
  multiplication, 163
  notation, 160, 171–172
  ordering, 167–169
  subtraction, 164–165
Real numbers, 199–256
  arithmetic, 199–200, 204
  completeness, 227–239
  definition, 199
  field, relation to ordered field, 221–227
  notation, 199, 226
  ordering, 209–211
  roots, 239–243
  sequence (*see* Sequence of rational numbers)
Real zero, 129–130
Reciprocal:
  of rational numbers, definition, 166
  of real numbers, 212–213
    definition, 212
Recursive definitions, 120–131
Reduction of fraction to lowest terms, definition, 162
Reflexiveness of equals, 10
Relation, 46–49
  definition, 47, 49
  function, 50
Relatively prime integers, definition, 151
Remainder, 150
Restriction of a function, definition, 55
Result of a count, definition, 99
Right component of ordered pair, 45
Roots of real numbers, 239–243

Second component of ordered pair, 118

Sentence, in logic, 18–24
Sequence, 117–119
  null, 194–199
  of rational numbers:
    arithmetic, 180–187
    Cantor, 187–194
    definition, 179
Single-element sets, 38–40
Single-valued function (*see* Function)
Square root of two, 177–179, 194–195, 201–203
Strict inequality, 94
  for integers, 140
  for rational numbers, 168
Subfield, 221–222
  definition, 221
Subset, 12–14
  definition, 13
  proper and improper, definition, 18
Subtraction:
  of integers, 142–143
  of natural numbers, 132
  of rational numbers, 164–165
Successor, definition, 77
Sum:
  definition, 107
  of integers, definition, 137
  of rational numbers, definition, 160
  of real numbers, definition, 200
  of sequences of rational numbers, definition, 180
Supremum, 229–238, 244–246
  definition, 229
Symbols for numbers and binary operations, 226–227
Symmetry of equals, 10
Synonym, definition, 5

Test pair for Cantor sequence, definition, 206
Transitiveness of equals, 10
Triangle inequality, 149
Truth table:
  for conjunction, 20
  for disjunction, 19
  for equivalence, 22
  for negation, 21

Union of sets, 27–28, 37
  definition, 28, 37
Unique Factorization Theorem, 154–156
Universal quantifier, 8
Unordered pair, 43
Upper bound, 244–246
  for subset of ordered field, 228–238
    definition, 228

Value of a function, 53
Variable, 7–9
  definition, 7
Venn diagrams, 16, 30
Void set, 16 (*see also* Empty set)

Weak inequality, 94
  for integers, 140
  for rational numbers, 168
Well-Ordering Principle (WOP), 96–98
  for nonnegative integers, 145
WOP (*see* Well-Ordering Principle)

Zero:
  definition, 76
  of polynomial function of real numbers, 129
Zero element:
  for a field, 215–216
  for integers, 138

Mathematics–Bestsellers

HANDBOOK OF MATHEMATICAL FUNCTIONS: with Formulas, Graphs, and Mathematical Tables, Edited by Milton Abramowitz and Irene A. Stegun. A classic resource for working with special functions, standard trig, and exponential logarithmic definitions and extensions, it features 29 sets of tables, some to as high as 20 places. 1046pp. 8 x 10 1/2. 0-486-61272-4

ABSTRACT AND CONCRETE CATEGORIES: The Joy of Cats, Jiri Adamek, Horst Herrlich, and George E. Strecker. This up-to-date introductory treatment employs category theory to explore the theory of structures. Its unique approach stresses concrete categories and presents a systematic view of factorization structures. Numerous examples. 1990 edition, updated 2004. 528pp. 6 1/8 x 9 1/4. 0-486-46934-4

MATHEMATICS: Its Content, Methods and Meaning, A. D. Aleksandrov, A. N. Kolmogorov, and M. A. Lavrent'ev. Major survey offers comprehensive, coherent discussions of analytic geometry, algebra, differential equations, calculus of variations, functions of a complex variable, prime numbers, linear and non-Euclidean geometry, topology, functional analysis, more. 1963 edition. 1120pp. 5 3/8 x 8 1/2. 0-486-40916-3

INTRODUCTION TO VECTORS AND TENSORS: Second Edition--Two Volumes Bound as One, Ray M. Bowen and C.-C. Wang. Convenient single-volume compilation of two texts offers both introduction and in-depth survey. Geared toward engineering and science students rather than mathematicians, it focuses on physics and engineering applications. 1976 edition. 560pp. 6 1/2 x 9 1/4. 0-486-46914-X

AN INTRODUCTION TO ORTHOGONAL POLYNOMIALS, Theodore S. Chihara. Concise introduction covers general elementary theory, including the representation theorem and distribution functions, continued fractions and chain sequences, the recurrence formula, special functions, and some specific systems. 1978 edition. 272pp. 5 3/8 x 8 1/2. 0-486-47929-3

ADVANCED MATHEMATICS FOR ENGINEERS AND SCIENTISTS, Paul DuChateau. This primary text and supplemental reference focuses on linear algebra, calculus, and ordinary differential equations. Additional topics include partial differential equations and approximation methods. Includes solved problems. 1992 edition. 400pp. 7 1/2 x 9 1/4. 0-486-47930-7

PARTIAL DIFFERENTIAL EQUATIONS FOR SCIENTISTS AND ENGINEERS, Stanley J. Farlow. Practical text shows how to formulate and solve partial differential equations. Coverage of diffusion-type problems, hyperbolic-type problems, elliptic-type problems, numerical and approximate methods. Solution guide available upon request. 1982 edition. 414pp. 6 1/8 x 9 1/4. 0-486-67620-X

VARIATIONAL PRINCIPLES AND FREE-BOUNDARY PROBLEMS, Avner Friedman. Advanced graduate-level text examines variational methods in partial differential equations and illustrates their applications to free-boundary problems. Features detailed statements of standard theory of elliptic and parabolic operators. 1982 edition. 720pp. 6 1/8 x 9 1/4. 0-486-47853-X

LINEAR ANALYSIS AND REPRESENTATION THEORY, Steven A. Gaal. Unified treatment covers topics from the theory of operators and operator algebras on Hilbert spaces; integration and representation theory for topological groups; and the theory of Lie algebras, Lie groups, and transform groups. 1973 edition. 704pp. 6 1/8 x 9 1/4. 0-486-47851-3

Browse over 9,000 books at www.doverpublications.com

A SURVEY OF INDUSTRIAL MATHEMATICS, Charles R. MacCluer. Students learn how to solve problems they'll encounter in their professional lives with this concise single-volume treatment. It employs MATLAB and other strategies to explore typical industrial problems. 2000 edition. 384pp. 5 3/8 x 8 1/2.   0-486-47702-9

NUMBER SYSTEMS AND THE FOUNDATIONS OF ANALYSIS, Elliott Mendelson. Geared toward undergraduate and beginning graduate students, this study explores natural numbers, integers, rational numbers, real numbers, and complex numbers. Numerous exercises and appendixes supplement the text. 1973 edition. 368pp. 5 3/8 x 8 1/2.   0-486-45792-3

A FIRST LOOK AT NUMERICAL FUNCTIONAL ANALYSIS, W. W. Sawyer. Text by renowned educator shows how problems in numerical analysis lead to concepts of functional analysis. Topics include Banach and Hilbert spaces, contraction mappings, convergence, differentiation and integration, and Euclidean space. 1978 edition. 208pp. 5 3/8 x 8 1/2.   0-486-47882-3

FRACTALS, CHAOS, POWER LAWS: Minutes from an Infinite Paradise, Manfred Schroeder. A fascinating exploration of the connections between chaos theory, physics, biology, and mathematics, this book abounds in award-winning computer graphics, optical illusions, and games that clarify memorable insights into self-similarity. 1992 edition. 448pp. 6 1/8 x 9 1/4.   0-486-47204-3

SET THEORY AND THE CONTINUUM PROBLEM, Raymond M. Smullyan and Melvin Fitting. A lucid, elegant, and complete survey of set theory, this three-part treatment explores axiomatic set theory, the consistency of the continuum hypothesis, and forcing and independence results. 1996 edition. 336pp. 6 x 9.   0-486-47484-4

DYNAMICAL SYSTEMS, Shlomo Sternberg. A pioneer in the field of dynamical systems discusses one-dimensional dynamics, differential equations, random walks, iterated function systems, symbolic dynamics, and Markov chains. Supplementary materials include PowerPoint slides and MATLAB exercises. 2010 edition. 272pp. 6 1/8 x 9 1/4.   0-486-47705-3

ORDINARY DIFFERENTIAL EQUATIONS, Morris Tenenbaum and Harry Pollard. Skillfully organized introductory text examines origin of differential equations, then defines basic terms and outlines general solution of a differential equation. Explores integrating factors; dilution and accretion problems; Laplace Transforms; Newton's Interpolation Formulas, more. 818pp. 5 3/8 x 8 1/2.   0-486-64940-7

MATROID THEORY, D. J. A. Welsh. Text by a noted expert describes standard examples and investigation results, using elementary proofs to develop basic matroid properties before advancing to a more sophisticated treatment. Includes numerous exercises. 1976 edition. 448pp. 5 3/8 x 8 1/2.   0-486-47439-9

THE CONCEPT OF A RIEMANN SURFACE, Hermann Weyl. This classic on the general history of functions combines function theory and geometry, forming the basis of the modern approach to analysis, geometry, and topology. 1955 edition. 208pp. 5 3/8 x 8 1/2.   0-486-47004-0

THE LAPLACE TRANSFORM, David Vernon Widder. This volume focuses on the Laplace and Stieltjes transforms, offering a highly theoretical treatment. Topics include fundamental formulas, the moment problem, monotonic functions, and Tauberian theorems. 1941 edition. 416pp. 5 3/8 x 8 1/2.   0-486-47755-X

Browse over 9,000 books at www.doverpublications.com

Mathematics–Algebra and Calculus

VECTOR CALCULUS, Peter Baxandall and Hans Liebeck. This introductory text offers a rigorous, comprehensive treatment. Classical theorems of vector calculus are amply illustrated with figures, worked examples, physical applications, and exercises with hints and answers. 1986 edition. 560pp. 5 3/8 x 8 1/2.   0-486-46620-5

ADVANCED CALCULUS: An Introduction to Classical Analysis, Louis Brand. A course in analysis that focuses on the functions of a real variable, this text introduces the basic concepts in their simplest setting and illustrates its teachings with numerous examples, theorems, and proofs. 1955 edition. 592pp. 5 3/8 x 8 1/2.   0-486-44548-8

ADVANCED CALCULUS, Avner Friedman. Intended for students who have already completed a one-year course in elementary calculus, this two-part treatment advances from functions of one variable to those of several variables. Solutions. 1971 edition. 432pp. 5 3/8 x 8 1/2.   0-486-45795-8

METHODS OF MATHEMATICS APPLIED TO CALCULUS, PROBABILITY, AND STATISTICS, Richard W. Hamming. This 4-part treatment begins with algebra and analytic geometry and proceeds to an exploration of the calculus of algebraic functions and transcendental functions and applications. 1985 edition. Includes 310 figures and 18 tables. 880pp. 6 1/2 x 9 1/4.   0-486-43945-3

BASIC ALGEBRA I: Second Edition, Nathan Jacobson. A classic text and standard reference for a generation, this volume covers all undergraduate algebra topics, including groups, rings, modules, Galois theory, polynomials, linear algebra, and associative algebra. 1985 edition. 528pp. 6 1/8 x 9 1/4.   0-486-47189-6

BASIC ALGEBRA II: Second Edition, Nathan Jacobson. This classic text and standard reference comprises all subjects of a first-year graduate-level course, including in-depth coverage of groups and polynomials and extensive use of categories and functors. 1989 edition. 704pp. 6 1/8 x 9 1/4.   0-486-47187-X

CALCULUS: An Intuitive and Physical Approach (Second Edition), Morris Kline. Application-oriented introduction relates the subject as closely as possible to science with explorations of the derivative; differentiation and integration of the powers of x; theorems on differentiation, antidifferentiation; the chain rule; trigonometric functions; more. Examples. 1967 edition. 960pp. 6 1/2 x 9 1/4.   0-486-40453-6

ABSTRACT ALGEBRA AND SOLUTION BY RADICALS, John E. Maxfield and Margaret W. Maxfield. Accessible advanced undergraduate-level text starts with groups, rings, fields, and polynomials and advances to Galois theory, radicals and roots of unity, and solution by radicals. Numerous examples, illustrations, exercises, appendixes. 1971 edition. 224pp. 6 1/8 x 9 1/4.   0-486-47723-1

AN INTRODUCTION TO THE THEORY OF LINEAR SPACES, Georgi E. Shilov. Translated by Richard A. Silverman. Introductory treatment offers a clear exposition of algebra, geometry, and analysis as parts of an integrated whole rather than separate subjects. Numerous examples illustrate many different fields, and problems include hints or answers. 1961 edition. 320pp. 5 3/8 x 8 1/2.   0-486-63070-6

LINEAR ALGEBRA, Georgi E. Shilov. Covers determinants, linear spaces, systems of linear equations, linear functions of a vector argument, coordinate transformations, the canonical form of the matrix of a linear operator, bilinear and quadratic forms, and more. 387pp. 5 3/8 x 8 1/2.   0-486-63518-X

Browse over 9,000 books at www.doverpublications.com

Mathematics–Probability and Statistics

BASIC PROBABILITY THEORY, Robert B. Ash. This text emphasizes the probabilistic way of thinking, rather than measure-theoretic concepts. Geared toward advanced undergraduates and graduate students, it features solutions to some of the problems. 1970 edition. 352pp. 5 3/8 x 8 1/2.   0-486-46628-0

PRINCIPLES OF STATISTICS, M. G. Bulmer. Concise description of classical statistics, from basic dice probabilities to modern regression analysis. Equal stress on theory and applications. Moderate difficulty; only basic calculus required. Includes problems with answers. 252pp. 5 5/8 x 8 1/4.   0-486-63760-3

OUTLINE OF BASIC STATISTICS: Dictionary and Formulas, John E. Freund and Frank J. Williams. Handy guide includes a 70-page outline of essential statistical formulas covering grouped and ungrouped data, finite populations, probability, and more, plus over 1,000 clear, concise definitions of statistical terms. 1966 edition. 208pp. 5 3/8 x 8 1/2.   0-486-47769-X

GOOD THINKING: The Foundations of Probability and Its Applications, Irving J. Good. This in-depth treatment of probability theory by a famous British statistician explores Keynesian principles and surveys such topics as Bayesian rationality, corroboration, hypothesis testing, and mathematical tools for induction and simplicity. 1983 edition. 352pp. 5 3/8 x 8 1/2.   0-486-47438-0

INTRODUCTION TO PROBABILITY THEORY WITH CONTEMPORARY APPLICATIONS, Lester L. Helms. Extensive discussions and clear examples, written in plain language, expose students to the rules and methods of probability. Exercises foster problem-solving skills, and all problems feature step-by-step solutions. 1997 edition. 368pp. 6 1/2 x 9 1/4.   0-486-47418-6

CHANCE, LUCK, AND STATISTICS, Horace C. Levinson. In simple, non-technical language, this volume explores the fundamentals governing chance and applies them to sports, government, and business. "Clear and lively ... remarkably accurate." – *Scientific Monthly.* 384pp. 5 3/8 x 8 1/2.   0-486-41997-5

FIFTY CHALLENGING PROBLEMS IN PROBABILITY WITH SOLUTIONS, Frederick Mosteller. Remarkable puzzlers, graded in difficulty, illustrate elementary and advanced aspects of probability. These problems were selected for originality, general interest, or because they demonstrate valuable techniques. Also includes detailed solutions. 88pp. 5 3/8 x 8 1/2.   0-486-65355-2

EXPERIMENTAL STATISTICS, Mary Gibbons Natrella. A handbook for those seeking engineering information and quantitative data for designing, developing, constructing, and testing equipment. Covers the planning of experiments, the analyzing of extreme-value data; and more. 1966 edition. Index. Includes 52 figures and 76 tables. 560pp. 8 3/8 x 11.   0-486-43937-2

STOCHASTIC MODELING: Analysis and Simulation, Barry L. Nelson. Coherent introduction to techniques also offers a guide to the mathematical, numerical, and simulation tools of systems analysis. Includes formulation of models, analysis, and interpretation of results. 1995 edition. 336pp. 6 1/8 x 9 1/4.   0-486-47770-3

INTRODUCTION TO BIOSTATISTICS: Second Edition, Robert R. Sokal and F. James Rohlf. Suitable for undergraduates with a minimal background in mathematics, this introduction ranges from descriptive statistics to fundamental distributions and the testing of hypotheses. Includes numerous worked-out problems and examples. 1987 edition. 384pp. 6 1/8 x 9 1/4.   0-486-46961-1

Browse over 9,000 books at www.doverpublications.com

Mathematics–History

THE WORKS OF ARCHIMEDES, Archimedes. Translated by Sir Thomas Heath. Complete works of ancient geometer feature such topics as the famous problems of the ratio of the areas of a cylinder and an inscribed sphere; the properties of conoids, spheroids, and spirals; more. 326pp. 5 3/8 x 8 1/2. 0-486-42084-1

THE HISTORICAL ROOTS OF ELEMENTARY MATHEMATICS, Lucas N. H. Bunt, Phillip S. Jones, and Jack D. Bedient. Exciting, hands-on approach to understanding fundamental underpinnings of modern arithmetic, algebra, geometry and number systems examines their origins in early Egyptian, Babylonian, and Greek sources. 336pp. 5 3/8 x 8 1/2. 0-486-25563-8

THE THIRTEEN BOOKS OF EUCLID'S ELEMENTS, Euclid. Contains complete English text of all 13 books of the Elements plus critical apparatus analyzing each definition, postulate, and proposition in great detail. Covers textual and linguistic matters; mathematical analyses of Euclid's ideas; classical, medieval, Renaissance and modern commentators; refutations, supports, extrapolations, reinterpretations and historical notes. 995 figures. Total of 1,425pp. All books 5 3/8 x 8 1/2.

Vol. I: 443pp.   0-486-60088-2
Vol. II: 464pp. 0-486-60089-0
Vol. III: 546pp. 0-486-60090-4

A HISTORY OF GREEK MATHEMATICS, Sir Thomas Heath. This authoritative two-volume set that covers the essentials of mathematics and features every landmark innovation and every important figure, including Euclid, Apollonius, and others. 5 3/8 x 8 1/2.

Vol. I: 461pp.   0-486-24073-8
Vol. II: 597pp. 0-486-24074-6

A MANUAL OF GREEK MATHEMATICS, Sir Thomas L. Heath. This concise but thorough history encompasses the enduring contributions of the ancient Greek mathematicians whose works form the basis of most modern mathematics. Discusses Pythagorean arithmetic, Plato, Euclid, more. 1931 edition. 576pp. 5 3/8 x 8 1/2.
0-486-43231-9

CHINESE MATHEMATICS IN THE THIRTEENTH CENTURY, Ulrich Libbrecht. An exploration of the 13th-century mathematician Ch'in, this fascinating book combines what is known of the mathematician's life with a history of his only extant work, the Shu-shu chiu-chang. 1973 edition. 592pp. 5 3/8 x 8 1/2.
0-486-44619-0

PHILOSOPHY OF MATHEMATICS AND DEDUCTIVE STRUCTURE IN EUCLID'S ELEMENTS, Ian Mueller. This text provides an understanding of the classical Greek conception of mathematics as expressed in Euclid's Elements. It focuses on philosophical, foundational, and logical questions and features helpful appendixes. 400pp. 6 1/2 x 9 1/4. 0-486-45300-6

BEYOND GEOMETRY: Classic Papers from Riemann to Einstein, Edited with an Introduction and Notes by Peter Pesic. This is the only English-language collection of these 8 accessible essays. They trace seminal ideas about the foundations of geometry that led to Einstein's general theory of relativity. 224pp. 6 1/8 x 9 1/4. 0-486-45350-2

HISTORY OF MATHEMATICS, David E. Smith. Two-volume history – from Egyptian papyri and medieval maps to modern graphs and diagrams. Non-technical chronological survey with thousands of biographical notes, critical evaluations, and contemporary opinions on over 1,100 mathematicians. 5 3/8 x 8 1/2.

Vol. I: 618pp.   0-486-20429-4
Vol. II: 736pp. 0-486-20430-8

Browse over 9,000 books at www.doverpublications.com

Physics

THEORETICAL NUCLEAR PHYSICS, John M. Blatt and Victor F. Weisskopf. An uncommonly clear and cogent investigation and correlation of key aspects of theoretical nuclear physics by leading experts: the nucleus, nuclear forces, nuclear spectroscopy, two-, three- and four-body problems, nuclear reactions, beta-decay and nuclear shell structure. 896pp. 5 3/8 x 8 1/2. 0-486-66827-4

QUANTUM THEORY, David Bohm. This advanced undergraduate-level text presents the quantum theory in terms of qualitative and imaginative concepts, followed by specific applications worked out in mathematical detail. 655pp. 5 3/8 x 8 1/2.
0-486-65969-0

ATOMIC PHYSICS AND HUMAN KNOWLEDGE, Niels Bohr. Articles and speeches by the Nobel Prize–winning physicist, dating from 1934 to 1958, offer philosophical explorations of the relevance of atomic physics to many areas of human endeavor. 1961 edition. 112pp. 5 3/8 x 8 1/2. 0-486-47928-5

COSMOLOGY, Hermann Bondi. A co-developer of the steady-state theory explores his conception of the expanding universe. This historic book was among the first to present cosmology as a separate branch of physics. 1961 edition. 192pp. 5 3/8 x 8 1/2.
0-486-47483-6

LECTURES ON QUANTUM MECHANICS, Paul A. M. Dirac. Four concise, brilliant lectures on mathematical methods in quantum mechanics from Nobel Prize-winning quantum pioneer build on idea of visualizing quantum theory through the use of classical mechanics. 96pp. 5 3/8 x 8 1/2. 0-486-41713-1

THE PRINCIPLE OF RELATIVITY, Albert Einstein and Frances A. Davis. Eleven papers that forged the general and special theories of relativity include seven papers by Einstein, two by Lorentz, and one each by Minkowski and Weyl. 1923 edition. 240pp. 5 3/8 x 8 1/2. 0-486-60081-5

PHYSICS OF WAVES, William C. Elmore and Mark A. Heald. Ideal as a classroom text or for individual study, this unique one-volume overview of classical wave theory covers wave phenomena of acoustics, optics, electromagnetic radiations, and more. 477pp. 5 3/8 x 8 1/2. 0-486-64926-1

THERMODYNAMICS, Enrico Fermi. In this classic of modern science, the Nobel Laureate presents a clear treatment of systems, the First and Second Laws of Thermodynamics, entropy, thermodynamic potentials, and much more. Calculus required. 160pp. 5 3/8 x 8 1/2. 0-486-60361-X

QUANTUM THEORY OF MANY-PARTICLE SYSTEMS, Alexander L. Fetter and John Dirk Walecka. Self-contained treatment of nonrelativistic many-particle systems discusses both formalism and applications in terms of ground-state (zero-temperature) formalism, finite-temperature formalism, canonical transformations, and applications to physical systems. 1971 edition. 640pp. 5 3/8 x 8 1/2. 0-486-42827-3

QUANTUM MECHANICS AND PATH INTEGRALS: Emended Edition, Richard P. Feynman and Albert R. Hibbs. Emended by Daniel F. Styer. The Nobel Prize–winning physicist presents unique insights into his theory and its applications. Feynman starts with fundamentals and advances to the perturbation method, quantum electrodynamics, and statistical mechanics. 1965 edition, emended in 2005. 384pp. 6 1/8 x 9 1/4. 0-486-47722-3

Browse over 9,000 books at www.doverpublications.com